UNINHABITABLE
A Case for Caution

C. S. GOLDSMITH

Goldstar Publications
2770 So. Maryland Parkway #416
Las Vegas, NV 89109
scgold711@aol.com

Cover designed by: Mandi Metzger
www.mandimetzger.com

Printed in the United States of America 2007
ISBN: 978-0-9795804-0-6

TABLE OF CONTENTS

My name is Ozymandias, King of Kings:
Look on my works, ye mighty, and despair!
Nothing besides remains. Round the decay
Of that colossal wreck, boundless and bare,
The lone and level sands stretch far away.
"Ozymandias"

- Percy B. Shelley (1792-1822)

To: Frederick Fleet and Reginald Lee:
Although trying to foresee any danger ahead,
failed to avert disaster.

-Lookouts for the R.M.S. Titanic

We are all passengers on the Titanic.

-Jack Foster, Irish philosopher

ACKNOWLEDGMENTS

To my mother, who instilled in me the love of nature and all animals. With thanks to Paul Thayer of Thayer Literary Services in Sarasota, Florida, and to Mike Sullivan, of Las Vegas, Nevada, for their work as editors and to Joyce Winters and Amanda Fox for their work on the manuscript. Without their skills, great help, and dedication, this book would never have been completed.

INTRODUCTION

This planet called Earth, our celestial home, is all we have--a relatively small terrestrial lifeboat, if you will, in the vast sea of the universe. Preserving and protecting this vessel, which supports us against all hazards, is of the utmost importance to our long-term survival on this biosphere.

Imagine life on Earth If something unexpected and virtually unknown reset the thermostat of the planet higher, instantly, and without warning. With only a relatively small increase in temperature of, say, an average of five to ten degrees, our world would change dramatically. We would see a drought so severe that it would create widespread crop failure and mass starvation. Incubated by a warmer climate, diseases such as malaria, cholera, and dung fever would become pandemic. El Nino-type storms would triple in intensity, with twenty to thirty major storms becoming the new annual norm. Those changes could cause billions of dollars' worth of damage and drive the insurance industry to the brink of ruin, creating a devastating ripple effect throughout the economy and the stock market. As the heat increased, the polar caps would melt, causing the oceans to rise by as much as twenty feet, erasing coastlines, flooding ports, and all but halting shipments of oil, natural gas, and other essentials. Petroleum reserves would diminish rapidly while prices skyrocketed. Without gasoline, millions of trucks and automobiles would sit abandoned, their drivers unable to get to work or deliver goods and services. Factory production would slow to a trickle. Billions of people would lose their jobs and billions more would be forced to flee from their homes as coastal flooding drove them inland. Homeless multitudes would wander the land in a desperate search for food, fresh water, and shelter. Before long, society would begin to disintegrate into a lawless and no doubt bloody struggle for survival. Now for the bad news--and unfortunately I'm not kidding.

Does this sound like the stuff of science fiction, perhaps the premise for a new disaster movie? It could be. A nightmare scenario like this could become all too real, because a potentially huge calamity is already gradually in the making. A forty-year, slow eco-bomb is ticking away, and we could see it explode with devastating effects within mere decades.

In a shorter span of time than has ever been imagined, civilization may find itself between a rock and a hard place--and something far worse. If the hundreds of scientists and climatologists are right, within the next ten years, we may pass a tipping point in the Earth's delicate climate balance. Then, in the decades that follow, we may face the shutdown of the Great Atlantic Conveyor belt, which will plunge the world into the next mini-ice age. Or we will finally pump enough CO_2 into the atmosphere to overheat the planet with an additional 7.4 degrees F. and tip the scales. What lies between the next ice age and overheating the planet is much more dangerous--a threat to our very survival that slumbers far below us, which is the real cause for concern.

On June 4, 2004, the Bush administration finally acknowledged for the first time what the rest of the world has been saying for the past twenty years: Global warming is caused by human activities and it will cause significant environmental changes in the next few decades. In other words, global warming is a real-world phenomenon, not just some half-baked scientific theory or the cries of some greenhouse-gas gloom-and-doomers.

Our government's reluctant admission is a very tardy one, to be sure, but I was glad to hear it, because I'm deeply concerned about our future. It comes so quickly, and once it's here, it can't be changed. What is done is done. We know the future holds adversities of all kinds--hurricanes, volcanic eruptions, and earthquakes, droughts and floods, urban violence, wars, and famine. A drought usually lasts just a few years, a hurricane spends its fury in only a few days, earthquakes create their destruction in minutes, and volcanic eruptions and wars eventually end. The uncontrolled, abnormal overheating of our planet would write a different story for our children, grandchildren, and even more distant generations if we don't take global warming seriously and start doing something about it now, before its damage cannot be undone.

I spent more than five years researching and writing this book. I am not a scientist. I simply followed the leads and researched the information that came to my attention years ago from a CIA report on global warming posted on the Net. Since then I have personally spoken with many scientists and climatologists. Then I connected the dots and made the case that is presented in this book. The position that I have tried my best to establish is simply that the world's civilization needs to exercise great caution before we hurl ourselves forward into a future that will add two billion more fossil fuel-burning vehicles to our roads and hundreds more coal-burning electrical plants, not to mention an additional three billion more mouths to feed.

Scientists, by their very nature, are usually introverts consumed by their research. They do not campaign or loudly vocalize their findings, especially if they are alarming. Their funding is usually obtained from the federal government, grants, or from academia, where those who control the purse strings insist that the protocol for releasing their information be confined to their small group of peers and to a few select scientific journals. Other than these accepted ways to unveil their research, scientists have no other forum that does not risk loss of funding or jeopardizing their careers. As we know, the government will place the best spin possible on any new scientific information that doesn't suit its political agenda, thus causing confusion, controversy, and political paralysis. Therefore, individual citizens must take it upon themselves to disseminate the facts and to clarify issues so that they can decide for themselves.

That's why I wrote this book. Here you will get the straightforward information that I was able to gather from reading, research, and discussions with a variety of scientists so that you can make up your own mind.

CHAPTER ONE

Not a Pretty Picture

As I wrote this book, the United States was experiencing one of the hottest years recorded in the past century. The previous winter saw virtually no snowfall in Anchorage, Alaska, traditionally one of the coldest places in the country. Trees and crops budded and bloomed earlier, changing the growing and harvest seasons. Droughts plagued most Western and many of the Midwestern states. Forest fires raged through the tinder-dry forests of the West in epic conflagrations that consumed hundreds of thousands of acres and destroyed houses and even small towns. Besides all that, the 2005 hurricane season produced a record number of storms, including two that left billions of dollars' worth of damage in their wake along the Gulf coast.

Is this normal or are these aberrations the warning signs of a larger danger?

During the course of my research I gradually assembled a picture of Mother Nature's current state of health, and it wasn't pretty. For instance:

- ☺ The present level of carbon monoxide in Earth's atmosphere will affect the weather and the environment for the next one hundred thousand years!

- ☺ Glaciers in the arctic and Alaska are receding at an alarming rate. A giant glacier that once resided at the edge of one Alaska town is now a thirty-minute drive away.

- ☺ Great chunks of ice the size of the State of Rhode Island are breaking off in the Antarctic.

☺ The ocean depths are getting warmer, changing the growth rates of plankton and algae and the feeding and breeding habits of many species of fish, birds, and mammals.

☺ Spring is arriving earlier in some parts of the world. When that happens, birds suffer because the insect populations that young birds feed on have peaked and moved on before they're hatched.

☺ Tree rings that show the age, growth, and health of trees are getting thinner every year, indicating distinct climatic changes.

☺ Even more alarming, perhaps, is the demise of whole species of frogs in Central and South America, a phenomenon that scientists first noted twenty-five years ago, when they suspected that something had gone terribly wrong in the ecosystems of the amphibians, but they couldn't nail down an answer. Because frogs are actually delicate creatures that have a semipermeable skin, scientists call them an "indicator species." Any environmental glitch causes them to die, which is like having a canary in a coal mine. If the canary dies, you know the miners are next.

In 2005, an international team of scientists finally provided an answer by documenting--for the first time--a direct connection between the extermination of about two-thirds of the 110 known species of harlequin frog and global warming. Increased heat by itself didn't cause all those frogs to die, but a paradoxical side effect of higher temperatures did. Global warming increases cloud cover in the tropical rainforests, which lowers daytime temperatures and encourages the growth of a cool-weather pathogen--the chytrid fungus, which attacks the sensitive skin of frogs.

The researchers also noted that the die-off of frogs and other species occurs about 80 percent of the time following warm years. They fear that more tropical species will become extinct since air temperatures in those regions rose three times as fast between 1975 and 2000 than for all the rest of the twentieth century.

The world as we know it, or knew it, is changing very rapidly, and the changes will have a major impact on everything and on all of us and on our children, and their children, perhaps with irreversible consequences that will unfold over the next twenty to fifty years. Although our federal government has finally decided to acknowledge what the scientific community has known for decades and the rest of the world knew twenty-five years ago, its position can be summed up by an official policy statement declaring it to be 100 percent committed to fossil fuels and to expanding capitalism and consumerism as rapidly as circumstances will allow. This declaration isn't too surprising when we know that President George W. Bush previously owned an oil company and that many of his friends and business associates in Texas are in the oil business. The Clinton administration wasn't much better. And we are all well aware of the power of the energy companies to exert influence on Washington. There are approximately One Trillion barrels of oil to be retrieved, if estimates are correct. At the predicted price of One Hundred Dollars per barrel, they have One Hundred Trillion reasons for not wanting us to change to a cleaner energy source.

Instead of focusing obsessively on protecting energy interests, our government leaders need to focus on that which sustains us all--a fragile, wounded, and beleaguered planet struggling to maintain a delicate ecological balance between optimal conditions for life to flourish and total disaster. They need to see that we are currently tinkering with all of our destinies like a small child playing with a light socket without any conception of what electricity can do. We are experimenting with and altering the basic conditions and natural mechanisms that allow us and every other living thing to exist on this planet.

The report from the 1999 Global Futures Project, a combined effort of the CIA, the National Intelligence Council (NIC), and NGOs (nongovernmental organizations) clearly states how global warming and the ecological decline of natural habitats will affect us in the future. So our elected leaders have definitely known since 1999 where we are headed and what the consequences of global warming would be, but they just chose to keep it to themselves and adopt a policy that would appease Big Energy and Big Business. The mantra is globalization at all costs.

In September 2002, sixty-five thousand people from around the world met in Johannesburg, South Africa, to discuss how to slow global warming and achieve sustainable development. The secretary of the conference stated plainly, "If we do nothing to change our current indiscriminate patterns of development, we will compromise the security of the Earth and the people, plants, and animals that live on it." Since the Rio Summit in 1992, the world's consumption of energy has risen 21 percent. Hydrocarbon emissions are up 15 percent worldwide, and the average year-round temperature of the Earth has risen by approximately one degree. In addition, the target date for feeding more of the world's hungry has been set back fifteen years, from 2015 to 2030. We won't meet that goal, either.

More bad news: Since the Rio Summit, carbon-based emissions have increased in the United States by more than 18 percent, causing concerned environmentalists to predict that one of every four mammals, including polar bears, could be extinct within thirty to fifty years. The primate population has declined by 50 percent since 1980. Three million people die each year from the effects of air pollution. Half the people on the planet have inadequate or polluted water supplies, and 2.2 million die annually from contaminated water.

Forty-five states in the U.S. are suffering under severe or extreme drought, because 2005 was one of the hottest years of the last century. In California, some regions are the driest recorded since 1883. The Colorado River is running at 17 percent below its normal flow, the lowest in one hundred fifty years of record-keeping. This, in turn, is affecting Lake Mead, which is down approximately eighty feet and may go completely dry within ten years. Scientists say this drought resembles the dust bowl disaster of the 1930s. On a scale of one to ten, the director of the Department of Agricultural says this three-year-old drought is a nine. To make matters worse, the abnormally dry conditions have spawned hundreds of wildfires, which have ravaged the West from California and Oregon to Colorado, Arizona, New Mexico, and Nevada, devastating millions of acres of timberland and wildlife habitats.

The loss of crops caused by drought affects food prices. Wheat harvests are down 14 percent, corn is down 7 percent, and soybean crops are down by 9 percent, resulting in a loss of more than $9 billion to agricultural communities in the U.S. All of this has now been scientifically proven to be directly linked to warming of the planet.

While many areas are suffering through a drought, other places are experiencing severe storms and harsh winters. Hurricanes, tornadoes, and other types of El Nino-related activity are wreaking havoc and costing billions of dollars in damage along the coastlines. Outside the U.S., many places like China and South America are reporting the hottest temperatures in more than a thousand years. The world may have entered a transitional climatic period. Global warming is here and its effects are now and we've only increased the average temperature by .8 degrees. What will it be like at 7.4 degrees hotter?

Fortunately, more politicians in more countries are looking closely at the facts. By February 2005, most governments and scientists had accepted global warming as a fact, and many people are now working to verify scientific tests that measure and monitor the changes and their effect on the planet. Scientists are also researching ways to deal with the excessive CO_2 in the atmosphere. In addition, most developed countries, including Russia, have signed the Kyoto Accord, which commits these countries to limiting their pollution. Fourteen states recently sued the U.S. government to do more about global warming and have taken steps to reduce emissions as outlined by the Kyoto Accord. California has just voted into law that greenhouse emissions must be below the accepted Kyoto protocol.

The world needs to develop a sense of urgency In the last decade about two and a half percent of the world's forests, mostly the tropical rain forests, have vanished. One third of the world's coral reefs are seriously damaged. Sixty percent of the world's oceans are now overfished. Some 2.8 billion people, or about a third of the Earth's population, live on less than $2 a day, adding to the problems of conservation and world hunger.

Can we let such things continue? We simply don't have a lot more time to waste.

Converging Calamities

Let's look ahead to the decades between 2020 and 2040. Here in this twenty-year window is where I believe we face our greatest risks. In those years, civilization will still be dependent upon oil and fossil fuels for energy to feed an ever increasing world population. We have already started down the hill toward the end of our oil reserves. By the middle of this century most of the oil that is economically accessible will be depleted. This will place us in a gasoline crisis. Will we have enough fuel to power our cars and the farm equipment that harvests the crops needed to feed the hungry multitudes - enough to fuel the growing armada of gigantic fishing trawlers, with their five-mile-wide gill nets?

If it becomes necessary to ration fuel, will our increased efforts to provide food for everyone play out the soil and fish out the oceans? Certainly many maritime species will not be able to replenish themselves and will become extinct the same way that too many land animals have in the past hundred years. To feed a population that is growing by eighty million people a year, we will see many changes, including specialized breeding farms with millions of animals being fattened for slaughter, international farm syndicates that own vast tracts of farmland for growing potatoes or corn or wheat, and orchards owned by large corporate farms that stretch for many miles to produce everything from oranges and apples to grapes and strawberries. Because all of these great enterprises will be dependent on oil, fresh water, and a favorably mild climate to produce their harvest, they will be placed in jeopardy.

Today 97 percent of the U.S. and European population are fed by the remaining 3 percent of farmers and fishermen, with all their goods delivered to distribution centers on a just-in-time basis by hundreds of thousands of farm workers and countless trucks. About all of this we could say, "So far, so good," because so far life has been very good indeed to most of us in the developed world. We have become lulled into a false sense of security by shopping at Wal-Mart super centers filled with acres of goods, enough to fulfill the needs and desires of every consumer and by grocery stores that also stock appliances, apparel, electronics, and thousands of other items, all in one place. Walk out your door, and within a few blocks in any direction is a store where you can buy almost anything you wish. We enjoy a system of consumerism that has grown to almost unimaginable proportions over the course of two hundred years, and it's a system that we trust implicitly. It seems reliable, resilient, and, above all, it works consistently on a nick-of-time basis. How could anything happen to it? The system will always be there, we believe, because it has always been there for us so far.

Is our belief in the system, like many of our other perceptions of safety, more vulnerable than we think? The truth is, life is not as secure as we think. In reality, what seems to be a well-oiled and perfectly trustworthy system that provides all that we need to survive could be disrupted more easily than you know. In fact, any disruption of the system, whether it's in production, processing, or distribution, could cause the whole machine to quickly stall or disintegrate. One exceptionally dreadful event broadcast by the national media, for instance, could inspire a hoarding frenzy that would strip store shelves in a matter of days. That happened right after the terrorist attacks of September 11, 2001, when many panicked citizens rushed to buy such things as guns, gasoline, gas cans, and television sets. The truth is that we have only seven days' worth of food in all the stores and the distribution pipeline at any given time. Any catastrophic event, either natural or man-made, that triggers a disruption in that food supply system would soon create hoarding and chaos. Such conditions, we may reasonably assume, could last from a few days or weeks to as long as a year or two before we could stabilize the situation. What if we had to deal with an event that was much more profound than an isolated glitch in the system? What if the disaster created a situation that couldn't be fixed easily or quickly? That is what we are facing with the issue of global warming. When we have tipped the ecological scales too far, then no easy fix will be possible. What happens then, when the climate is too hot to grow food and when the crops wither and die in the fields? How do we put water back into the aquifers once they have been pumped dry? What do we do when the polar ice caps melt and the oceans rise twenty feet over the land? How do we stop a drought that will not go away because temperatures are on average seven to ten degrees hotter? How do we replenish the fish populations when they're on the verge of extinction because of rising ocean temperatures which prevent spawning, and because of overfishing that decimates the population? How do we restore the rainforests that help to purify the air we need to breathe once they have been cleared or burned or logged out? How do we replace the oil needed to run a global gas-guzzling economy?

The answer to all these questions, of course, is that we can't. We are in fact powerless to stop any natural calamity once it has started. The natural world will seek its own corrective measures to regain its balance, as it has always done, but that may take more time than Mankind has to spend.

All the tragedies of the twentieth century combined with those of every other period of recorded history would pale in contrast with the suffering and death the world would witness in the wake of a global ecological meltdown that we have unfortunately initiated ourselves. An Armageddon scenario as horrific as any depicted in the Book of Revelation will play out if we do not do something to control global warming, if we cannot find a way to reduce our world population to sustainable numbers, if we don't stop producing hundreds of millions more fossil fuel-burning vehicles, and if we don't do more to protect our oceans, fisheries, forests, farmlands, and fresh water resources. Precious little time remains for us to find and implement the solutions needed to protect what we have.

We are fast approaching a fateful crossroads. When we get there, then all that we have accomplished in the last five thousand years, all that we have now, and all that we may ever have in the future could be lost forever.

CHAPTER TWO

The Permian Parallel

As bad as conditions are now in our environment, they could get worse--a lot worse. To learn more about the consequences of ecological change, we can take a look at the history of our planet. Let's begin with the remote past and go all the way back to the Permian period, which began 248 million years ago.

More than forty million years before the Age of Dinosaurs even began in the Triassic period, the Permian period teemed with life--plants, fish, insects, amphibians, reptiles, and mammals. At one point, though, 95 percent of all living things were dashed to extinction, including the plant life and almost everything in the oceans, and the Earth came close to becoming a lifeless rock revolving around the sun.

The cause of this worldwide extermination was a mystery that puzzled scientists for decades. Speculation continued until a geologist found some convincing evidence preserved in the rocks and ash in Antarctica. Signs there indicated that a large meteor struck the Earth off the coast of Australia with such force that it destabilized the tectonic plates on the other side of the world in Siberia. That set off a chain of volcanic cracks in the Earth's surface that spewed billions of tons of molten ash and carbon dioxide (CO_2) into the atmosphere. Evidence of these volcanic eruptions was not easy to find, because the volcanic cracks filled in and became covered with earth and plants and the higher formations of lava eventually eroded away.

At first scientists thought the mass extinction was caused by a large meteor, because of the fractured quartz found in Antarctica. Then they realized that the real problem was caused by all the CO_2 disgorged by the volcanic eruptions, which increased the temperature of the atmosphere enough to warm the surface of the oceans. Within a short time, all the warm surface sea water percolated down to the ocean floor, which then triggered the release of vast methane hydrate deposits located just below the ocean floor. This global methane disassociation immediately accelerated the warming process and soon doubled the average temperature increase of the planet by more than ten degrees on average warmer than before the methane release, killing off most of the plants and then the plant-eating animals. Heat and other gases helped to deplete the oceans of oxygen, a condition known as anoxia, destroying most of the sea life as well. Millions of years would pass before new forms of plants and animals appeared once again on Earth and eighty million years passed before the planet was replenished, this time with dinosaurs at the top of the food chain.

Methane hydrate gas is particularly pernicious because it is fifty-six times more powerful a greenhouse gas than CO_2. Notably, carbon dioxide is the key that releases the genie from the ecological bottle. Internal combustion engines and industrial facilities are currently replicating the same effects created by all that volcanic activity of 248 million years ago by pumping approximately 57 billion tons of CO_2 into the atmosphere each year. Unfortunately, the U.S. government does not classify CO_2 as a pollutant, and it does not regulate its emission. If we do nothing, which is exactly what we have been doing for the past 25 years, the additional CO_2 in the upper atmosphere will increase temperatures around the globe by 3.5 to 7.4 degrees F. in the next thirty to seventy-five years. Even worse, if enough warm surface ocean water reaches the ocean floor to release the methane deposits, the world's temperature could spike by 5.5 to 10 degrees higher. History has taught us that a 10 degree spike in temperature will decimate life on this planet.

Other computer models (nearly two dozen of them) all suggest that the average yearly global temperature will rise by 3 to 7.4 degrees above present temperatures by 2100, according to

Gerald Meehl, a senior scientist at the National Center for Atmospheric Research. That is significant because each degree of increased temperature causes harvests around the world to decline by 10 percent. That, in turn, will greatly increase the difficulty of feeding the nine billion people that we're planning to have in that period of time.

A temperature increase of 3.5 to 7 degrees could conceivably liberate the world's deep-ocean methane deposits, which would act as a supercharger for atmospheric warming because of methane's unique ability to retain heat. Unlike CO_2, which remains in the atmosphere for hundreds of years, methane dissipates in a decade or two. But while it is present, it acts like a magnifying glass that concentrates heat and can ratchet up the temperatures by 5 to 10 degrees or more. I don't think any scientist believes that our planet could handle that kind of heat without losing most of the living organisms on it. At the very least, such a condition would fundamentally change everything about our present way of life to the point that we would not recognize it. Try to envision a planet with no trees, lots of deserts, and very little water.

Geologists tell us that this temperature spike from methane has happened five or six times before because of meteors, earthquakes, volcanic activity, or shifts of the Earth's temperatures that causes glacier build-up and reduces pressure on the methane causing disassociation. It happened as recently as fifty-five thousand years ago, although the effects were not as catastrophic as those of the Permian period. Now, for the first time in human history, our own activity has started the CO_2-methane release cycle and placed us squarely in harm's way. Beneath the oceans and the polar ice, slumbers a deadly behemoth in the form of a colorless, odorless gas that has the potential to destroy us and most of the life on the planet. That is the awesome power of methane hydrate.

The power of the Earth's reservoirs of trapped gases can be seen in microcosm in places like Lake Nyos in Africa, where high concentrations of CO_2 and methane bubble up from deep springs under the floor of the lakes. Water pressure keeps the CO_2 and methane gas contained at the bottom, but it continues to build and build slowly. Then, every thousand years or so, a giant explosion of CO_2 (and carbon monoxide) bursts like champagne from a shaken bottle and creates a tidal wave that races toward the shore. The freed gases also form a deadly white cloud of lethal carbon monoxide gas that spreads across the land and kills every living thing in its path. In the last instance, the eighteen hundred people living around Lake Nyos died in their sleep. Ninety-eight percent of the people and animals within one and a half miles of the lake were killed.

This phenomenon, known as "lake turnover," can be triggered prematurely by a geologic event like an earthquake, a landslide, a volcanic eruption, or a meteor strike. Scientists have found much larger lakes, including those that have millions of people living around them that have the same slow buildup of potentially lethal gas. Core samples have shown that without a triggering event, the gases will remain stable for thousands or even tens of thousands of years, but eventually they will build up enough pressure to erupt naturally, causing a terrible toll in death and destruction.

Similar turnover events can occur in the oceans, with enormous reservoirs of methane and methane hydrate, held in check by tremendous pressure and extremely cold temperatures. These temperatures freeze the gases into an icy permafrost. Any significant change in that temperature or pressure could release it and radically change our climate in short order. For instance, the onset of another ice age, caused by the slowing or shutting down of the Atlantic conveyor, would lower the level of the oceans by creating vast storehouses of ocean water in the form of massive glaciers that hold trillions of gallons of seawater. When ocean levels drop, the pressure that held the methane in check is removed and the trapped deposits of methane gas explode into the atmosphere, creating a greenhouse effect that heats up the planet very quickly. In time, the glaciers melt and return water to the ocean, increasing the water pressure to the point of sealing in the methane and again re-establishing the equilibrium that stops the whole process. The methane gases dissipate in the atmosphere over twenty years or so, and the climate returns to normal in what is known as a negative-feedback, climate-control system. Methane hydrate acts as the Earth's thermostat, raising temperatures when it gets too cold. The great Atlantic conveyor belt is the counterbalance that, when shut down, starts a cooling phase.

Long after the end of the Permian period, a major event abruptly ended the earthly reign of the dinosaurs sixty-five million years ago. This violent incident was probably more cataclysmic than had been previously thought. All the destruction began when a large meteor slammed into the Pacific Ocean near the Yucatan Peninsula. The impact spewed millions of metric tons of dirt and ash into the atmosphere, covering the whole globe with a dark shroud that blocked out the sun and sent a three-mile-high tidal wave sweeping across the ocean and far inland.

Furthermore, soot rings found at the threshold of the Paleocene epoch, beginning 65 million years ago, strongly suggest that the dinosaurs and other life forms had to contend with another lethal problem in their desperate bid to survive the catastrophe. A massive methane disassociation caused by the rupture of the ocean floor sent trillions of tons of methane into the atmosphere. Within moments of the impact, red-hot billy balls--molten globs of the fragmented meteor–may have ignited the freed methane and literally turned the atmosphere into liquid fire. This destroyed most of the plant and animal life and spelled doom for the dinosaurs.

After forty years of research, scientists found that Carbon-12 was the real clue to unraveling the mystery of what caused the mass extinction that occurred. Meteors and volcanoes have caused mass extinctions, but when these catastrophic events are combined with a massive methane release, they have the potential to destroy 95 percent of all life. We should be very careful about monitoring the level of Carbon-12 and methane in our atmosphere, because too much of it could mean that a significant methane disassociation is beginning.

The Pleistocene Surprise

Every ice age is followed by a period with a relatively warm climate, which is what we have enjoyed for the last ten thousand years during the latter part of the Pleistocene epoch. That just happens to coincide with the rise of homo sapiens and modern man to the top of the food chain. If we had not enjoyed the narrow parameters of an optimum climate, we could still be chipping stone tools instead of surfing the Internet. Human life requires an Earth that rigorously maintains a fairly precise ecological balance.

We are now on the threshold of a new cycle, but instead of cooling, this time the Earth is becoming warmer. That's because we have tampered with the global thermostat by increasing the level of CO_2 in the biosphere, which has been unnaturally higher since the start of the industrial age in the mid-1800s and which has accelerated dramatically since the number of automobiles in use reached half a billion in the 1960s. Raise the level of CO_2, and you reset the Earth's thermostat to a higher temperature and create a greenhouse effect. That makes an already warm planet even warmer and too warm is not good.

We can see what happens with a build-up of pressurized CO_2 by looking at what has happened at Lake Nyos, where almost every living thing was wiped out by the release of vast quantities of gasses, primarily CO_2, which was held in place while building up pressure at the bottom of the lake. A different but equally dangerous situation occurs in the vast oceans of the world, except that there the critical gas is primarily methane, one of the world's most powerful greenhouse gases -- a massive methane release could have devastating consequences for all living things. The oceans show signs of enormous discharges of methane gas as recently as fifteen thousand years ago, which corresponds to the time when the last ice age ended.

We are not immune to this natural overturn and conditions now aren't much different from those of the Permian period. In fact, we may be only 3.5 to 7.5 degrees F. away from destabilizing the vast deposits of methane hydrate that lie just below the continental shelf off the coast of California and Oregon, in the Gulf of Mexico near Florida, Louisiana, and Texas, and elsewhere around the world. Methane reservoirs are usually found at a depth of five hundred meters or more below the ocean floor, although they lie comparatively close to the surface in some places like the Gulf of Mexico. Beneath the permafrost of the arctic, under the glaciers of Iceland, and below Greenland's ice sheets, the methane deposits are much closer to the surface--a mere 120 meters down. Since melting is occurring at an alarming rate in this part of the world, this could very well be the place where we draw our first battle lines in the war against global warming. We must also maintain a close watch on the relatively shallow waters of the Gulf of Mexico, where the temperature has already increased by more than three degrees, the Blake's Ridge area of Louisiana, Florida, and coastal California.

The defrosting of the Earth's icebound areas is of particular concern because the warming process is being accelerated. Normally, the polar ice caps and the vast sheets of arctic and antarctic ice act like a mirror that reflects much of the Sun's heat back into space. But as the ice melts, more of the much darker seawater is exposed, and then more of the Sun's energy is absorbed, increasing the temperature and causing even more ice to melt. This is one of the negative feedback loops of global warming that is hastening the process--and inching the world closer to a disastrous discharge of methane gas.

How much methane are we talking about? A truly vast amount. The Earth contains twice as much methane as all other energy deposits combined, including oil, coal, and all natural gases--an unimaginable 10 percent of the entire biomass of the Earth. A huge portion of it is just waiting for the appropriate amount of warmth to reach the bottom of the ocean or a drop in ocean pressure to set it free.

With the release of that much methane, the world would reel from a very quick and dramatic spike in the average world temperature. If the heat increased enough, human habitation -- and that of most plants, animals, and fish would come to an abrupt end, just as it did for almost all the species of the Permian period and for most of the species larger than an alligator during the Jurassic.

Even without a sudden spike in the Earth's temperature, the gradual warming of the planet, even over hundreds of years, could ultimately spell disaster for humanity. When we think about an ice age coming to an end, we see it happening in the creeping, glacial pace of geological time. Some scientists who study climate change, though, are seriously considering a phenomenon that is much more alarming -- the possibility of abrupt climate change. In some cases, they tell us, even complex systems like the Earth's atmosphere need only a brief transitional period in order to shift from one climatic state to another. Analogous to this concept is a pot of water heating on the stove. The water gradually grows hotter and hotter and then suddenly turns into steam.

Geoscientist Richard Alley of Pennsylvania State University studied ice core samples he took from Greenland in the 1990s and made a shocking discovery. He found that the last ice age ended in much the same way as the water-to-steam event, not with painful slowness but suddenly, with the whole planet warming in just three years. In terms of geological time, that's a mere blink of an eye. As a result of the study, Alley concluded that "There are thresholds one crosses, and [then] change [moves] a lot faster." Usually, he said, "climate responds as if it's being controlled by a dial, but occasionally it acts as if it's controlled by a switch." After studying rapid climate change in the arctic, Laurence Smith, a UCLA associate professor of geography, agreed with this idea. "We face the possibility of abrupt changes," he said, "that are economically and socially frightening."

This surprising evidence about the last ice age in the Pleistocene holds a chilling warning for us today. If the climate of Earth can change that quickly during the natural course of events, without any negative human influences, then what can prevent our reaching the boiling point by doubling the amount increasing the output of CO_2 into the air in the next two decades and by warming our oceans' surfaces by three to five degrees or more? Unless chemistry is inaccurate, one thing is certain, there will be a reaction.

What the Scientists Are Saying

Certainly the abundance of methane hydrate in coastal waters of relatively shallow depth could have a significant impact on atmospheric composition and global warming. Increasing the temperature of the oceans by one-half to one degree C. has the potential of creating an enormous release of methane into the atmosphere, which would result in further global warming (MacDonald, 1990; Paul, et al, 1991).

Because natural gas hydrate is metastable (that is, having only a slight margin of stability) and is also affected by changes in pressure and temperature, any release of methane is an agent that could affect oceanic and atmospheric chemistry and ultimately the global climate. Thus, these ramifications of natural gas events all have potential effects on human welfare. (K.A. Kvenvolden. U.S. Geological Survey.)

Naturally occurring gas hydrate was first discovered in the 1960s by Russian scientists who were exploring for gas in the permafrost region of the northern Soviet Union. By 1980, it had also been found on the continental slope in the Middle American Trench in the Gulf of Mexico near Guatemala and in the U.S. in Texas and Louisiana in the Blake's Ridge Range. As the thermal signature of global warming penetrates the ocean, a precise knowledge of the stability of gas hydrates will be required to assess the risks of decomposition of these reservoirs (Levitus, et al, 2000).

The gas hydrate reservoirs in ocean sediments have significant implications for the climate because of the enormous amount of methane situated there and the strong greenhouse warming potential of methane in the atmosphere. Methane absorbs energy at wavelengths that are different from other greenhouse gases, so a small addition of methane can have profound effects. If a mass of methane is released into the atmosphere, it will have an immediate greenhouse impact that will slowly decrease as the methane is oxidized into carbon dioxide in the air over a twenty-year period. A unit mass of methane introduced into the atmosphere would have fifty-six times the warming effect of an identical unit of carbon dioxide (William P. Dillion, Woods Hole, Mass.).

Atmospheric warming has increased the global surface temperature by roughly eight-tenths of a percent over the last hundred years. This warming is probably being transferred to the oceans in a manner comparable to "chemical tracers," which means that warmed surface water can be expected to circulate down to the depths of the more shallow gas hydrate deposits within several decades and in a shorter time in specific cases in the Gulf of Mexico where warm water currents sometimes sweep through the region and in northern California. Active disassociation of gas hydrates at the sea floor has been observed. Some of this activity has been related to identifiable water changes, and present atmospheric warming may be leading to hydrate disassociation that would reinforce the warming trend (William P. Dillion, Woods Hole, Mass.; Michael D. Max, Marine Desalinization Systems, L.L.C.).

In the long view, methane from gas hydrate may have had a stabilizing influence on the global climate. When the Earth cools at the beginning of an ice age, the expansion of glaciers binds up seawater in vast continental ice sheets, which causes the sea level to drop, thereby reducing pressure on the hydrates in the sea floor and causing disassociation and release. Such liberation of methane could increase the greenhouse effect and cause global warming. Thus, we may speculate that methane hydrate might be part of a great negative feedback mechanism that leads to the stabilization of the Earth's temperature. Methane is 56 times more effective as a greenhouse gas than CO_2 gas; furthermore, a methane release from hydrates could create a positive feedback for a global warming temperature increase of 5 degrees more which would result in hydrate instability and more methane releases into the atmosphere, increasing the temperature further. (Peter Wellsburg and John Parks, Department of Earth Sciences, University of Bristol, U.K.).

The Blake's Ridge Range may contain eighteen trillion cubic meters of gas hydrates plus another seventy to eighty trillion cubic meters of gas off the coast of Texas and Louisiana beneath the waters of the Gulf of Mexico.

The data suggest that a time lag occurs between the warming of the bottom seawater and the warming of the surface water and the atmosphere; thus, this model implies that hydrate disassociation, which would be the consequence of bottom-water warming, could lead to the catastrophic injection of methane into the atmosphere and the acceleration of greenhouse warming. (Kennett, et al, 2000).

Severinghaus and Brook (1999) suggest that warming occurs a few decades before the peak in atmospheric methane volumes, implying that warming causes a methane release and may not be a consequence of it.

A prominent negative shift in global carbonate concentrations during the late Paleocene warming trend, which occurred 55.5 million years ago, as recorded in sediments worldwide, indicates a catastrophic infusion of 812 C; enriched carbon, probably from a methane source. A rapid warming of bottom seawater by four degrees C. at this time could have abruptly altered the sediment. A thermal increase at ocean floor levels encourages the catastrophic release of methane from gas hydrates. Kennett and Stotl, 1991; Kock, et al, 1992; Bralower, et al, 1995; Dickens, et al.

Terrestrial or above ground methane is responsible for 70 percent of all current methane released into the atmosphere and is only 2 to 3 percent of the total methane contained in deep-sea sediments which accounts for 10 percent of the entire biomass on the Earth.

CHAPTER THREE

Red Flags

Controversy will swirl on any issue. Not everyone agrees on where to lay the blame for global warming. Some scientists say it's part of a natural process, not a man-made condition but just a cyclical temperature fluctuation; however, according to the overwhelming evidence now in, the majority of the world's scientists and climatologists say the problem is caused by humans and their activities. "There is no doubt that the climate is changing, and humans are responsible," says Kevin Trenberth, the head of the climate-analysis section at the National Center for Atmospheric Research (NCAR) in Boulder, Colorado. Whether the problem is man-made or natural--or a combination of both -- hardly matters now. The fact remains that the Earth is getting warmer, and the end results for the world will be the same in thirty years. Half of all the pollutants already in the atmosphere will still be there a hundred thousand years from now, even if all the emissions stopped today.

We have recently experienced nine of the warmest years on record. The warmest year ever recorded was 1998, with 2002, 2003, and 2004 not far behind, and the global average temperature in 2005 was thc warmest yet. Thanks to tree rings and ice core samples, which indicate atmospheric temperatures for thousands of years, we know that the Earth is approximately one degree warmer than the historical average of fifty-seven degrees and that it is becoming consistently hotter. According to estimates made by NASA scientists, our world hasn't been this hot for ten thousand years. The increasing heat has caused four hundred thousand square miles of Arctic ice to melt in the last thirty years--an area about the size of Texas--which has further accelerated global warming, according to the Arctic Climate Impact Assessment research. Glaciers are receding, too. Spring is arriving fourteen to seventeen days sooner, and growing seasons are changing.

Furthermore, we have seen a 100 percent increase in the intensity and duration of hurricanes and tropical storms since the 1970s, according to the National Climatic Data Center. The year 2005 went down as the worst hurricane season on record, joining 2004 as one of the most violent ever. In addition, El Nino-type storms have caused ten times more damage in the last decade than in all of recorded history. Changes in the weather and climatic change may be two different phenomena, as scientists are quick to point out, but the frequent occurrence of so-called once-in-a-lifetime storms in recent years has even skeptical members of the scientific community wondering if something terribly amiss is underway.

In a study of people all over the world, from the Great Barrier Reef to the forests of China and the plains of Africa, researchers have documented changes in the climate and its effect on the environment. In the United States, Las Vegas, Nevada, is suffering its worst drought in twenty years. Besides that, Las Vegas used to have cold winters, but now it is much warmer year-round. The U.S. has also had less rainfall in recent years than the average of the last hundred years. Obviously, things are changing rapidly. The debate over the cause of global warming is over now. Quantitative research has given us a definitive answer: global warming is here and now and it is caused by human activities. The only questions that remain are how we can reduce and stop what is contributing to the warming trend and how we can adapt to the environment of the future.

Significantly, the denial by our governmental leaders that the problem has been caused by humans has created a delayed response to the problem. Washington refused to sign the Kyoto Treaty and the Johannesburg Treaty. Although our leaders have finally acknowledged the problem of global warming, they still have no plans to abandon the use of fossil fuels. To make matters worse, the government rescinded the EPA's right to control the carbon monoxide emissions of older coal-burning plants so that facilities that are twenty to fifty years old could get permits to operate without adhering to the stricter EPA emission standards and continue to pour tons of pollutants into the air. In short, the government has done too little to protect our future for the sake of placating Big Energy interests.

Major U.S. energy companies have certainly had a friend in the White House with George W. Bush. Economic expansion and political favoritism for the oil companies under President Bush made more money for Big Oil in the first five years of this century than in the previous decade. Exxon Mobil alone made 64 billion in profits in 2005. The president supports oil interests and he has friends like those who control the Haliburton company, which was awarded a $52 billion contract to sell oil for $1.65 a gallon when they can buy it themselves in Iraq for thirty-five cents. Meanwhile, the U.S. is more than a half trillion dollars in debt and counting, which means that each American family now owes more than $55,000 in national debt, while the value of the dollar is sinking. Who do you think is going to bail us out and balance the books? Our children and grandchildren will, if they can, as taxpayers, and they may have to do it by bankrupting programs like Social Security and Medicare.

If you're waiting for the government to save us, think again very carefully, because our elected officials don't care about slowing down the economic gravy train if it means losing support from large corporations or special-interest groups. Even though they may see the train derailing in the future, the prevailing attitude is to let it go full steam ahead and hope for a miracle; besides, the next administration will have to deal with those problems.

Meanwhile, the temperature of our atmosphere will continue to rise, and all the ramifications of the warming trend will remain, just as surely as the sun rises in the East and sets in the West.

New Evidence

If you don't think that we have been piling up environmental debt the way the federal government has been escalating our fiscal debt, consider the results of a worldwide investigation completed recently by the National Geographic Society, one of the world's foremost authorities on nature and the environment. They found that CO_2 levels in the atmosphere are higher now than they have been in hundreds of thousands of years and that the world is .8 degrees hotter on average. This has already dried up rivers, created deserts, melted glaciers, and raised the sea level. By the end of this century, when the world could be five to seven degrees hotter, the imbalance could trigger a rapid climate change at some point.

Red flags are apparent everywhere. For example, when President Taft visited Glacier National Park in 1910, he counted 150 glaciers there. Now less than thirty remain, and those have shrunken by more than two-thirds. By 2030, the U.S. Geological Survey predicts, no glaciers will be left there at all. Mountain snowcaps around the world are melting; since 1912, the snows of Kilimanjaro have receded by 80 percent. Former Vice President Al Gore, in his film *An Inconvenient Truth*, predicts that Mount Kilimanjaro will lose all its snow within twenty years. In the mountains of India, the snow is retreating so rapidly that officials believe it will be gone completely in the central and eastern Himalayas by 2035. Another estimate says that the snowcaps in the mountains of Peru will be gone by 2100, leaving the population that depends on them for drinking water, farming, and electricity, high and dry. In Europe, scientists predict that most of the snow will be gone from the Alps within fifty years. The ice in the Arctic Sea has thinned by more than 10 percent in the last thirty years, and the Arctic spring ice thaw now occurs nine days earlier than it did a century ago.

In addition, the seas of the world have risen from three to eight inches in the last hundred years. Coastal scientists believe that every inch of rise in sea level could result in the loss of eight feet of beach by erosion. More than a hundred million people live within three feet of the ocean, making them extremely vulnerable. Furthermore, saltwater intrusion in freshwater aquifers would threaten drinking water supplies and agriculture in places like Egypt, where water is required for irrigation.

If the West Atlantic ice sheet broke off as the Larson ice shelf did in 2002, the sea level would rise by more than twenty feet. Even if that ice shelf remains intact through this century, scientists predict an increase of sea level of eight to thirty-five inches by the end of the century. The higher estimate would be nothing short of catastrophic. If the ice melts in Antarctica, the rise in sea level would be an unbelievable 235 feet higher than today.

A significant change in oceanic currents, especially those of the great Atlantic conveyor belt, could be just as bad. These vast currents within the oceans work like great conveyor belts that carry warm water from the equator to the poles, bringing either warmer or cooler weather to the peripheral land masses. Warmer temperatures are affected by the salinity of ocean water. Fresh water from melting glaciers reduces the salt in the oceans causing the Great Atlantic conveyor belt to descend further in the cooler ocean regions, slowing it down and cooling temperatures along its path. The greater the temperatures change, the more it affects our weather. The greater the temperatures change, the more it affects our weather. With water temperature rising, at some point the increased warmth could trigger a slowdown of the Great Atlantic conveyor belt. Some studies suggest that this current has already slowed by fifteen to 20 percent. Global warming could even shut it down completely, which would cause an abrupt climate change. Robert Gagosian of the Woods Hole Oceanographic Institution believes that too much change in ocean temperature and salinity could slow or shut down the Great Atlantic Conveyor belt and cause a dramatic climate shift in less than a decade. When that happens we will confront a nightmare similar to the one in the film *The Day after Tomorrow*, which depicted giant arctic storms that suddenly changed the world's climate and caused cataclysmic destruction and a horrendous loss of life.

Besides the oceans of our planet, other keys to the healthy life of our environment are three crucial atmospheric gases -- CO_2, methane, and nitrous oxide. Collectively acting like a global thermostat, these gases keep the Earth temperate and hospitable. Without an intricate balance of them in the atmosphere, our world would be just a cold, uninhabitable rock. According to NOAA, this balance remained optimal for thousands of years until about 1850, when the levels of all three gases began to increase in a trend that spiked sharply after 1950. This change exactly parallels our increasing use of fossil fuels from the middle of the nineteenth century onward. Obviously, our activities have significantly affected the atmospheric mixture of these gases.

We are forced to conclude that filling the atmosphere with even more of these gases during the next century will have a drastic effect on our climate. At some point, if temperatures continue to rise, all life on Earth will have no place to hide from the effects of a hotter planet. That unhappy day may not be long in coming; we may see major climatic changes in a mere twenty to fifty years.

Does Anybody Care?

I launched my work on this book in 1999, when I read a report on the Internet by the CIA-Global Futures Project. Now we have transitioned into the brave new world of the twenty-first century. Fortunately, the whole world, save for a handful of skeptics, now agrees that we have a global warming problem. Until May 2002, many scientists and political leaders, especially those in the U.S., denied that it was caused by humans or that it was a problem at all. Had we had the wisdom to implement the Kyoto protocol twenty years ago, we would have come a long way in confronting the problem. But the U.S. failed to sign the Kyoto Accord and get on board, because that would come with a price tag of billions of dollars. I wonder, though, what the cost will be if we don't face the challenge of global warming. Trillions of dollars is a likely estimate of the price we'll pay.

Russia, to its credit, has signed the Kyoto agreement, and many countries and international corporations are taking steps to reduce the pollution that contributes to global warming. These efforts may be a case of too little, too late, but only time will tell. Scientific studies are being issued almost every day with concrete evidence that the world's climate is changing rapidly. Note the killer hurricanes in the Gulf of Mexico and the destruction in New Orleans and the reduced rainfall in the western U.S. and the Andes Mountains. Consider the diminishing glaciers, where the spring thaw feeds the streams and irrigates the farms, and the loss of a habitat for polar bears. Inspect the satellite photos of Antarctica, which show a significant reduction in the polar ice cap, even to the untrained eye.

That's not all. We are now on the downhill side of retrieving fossil fuels economically. Geologists have determined that we reached our peak of oil extraction in 2004. No more major oil fields are being found. We're consuming two barrels of oil for each new barrel discovered. The International Energy Agency, the world's energy watchdog, has guessed that oil demand will rise from 85 million barrels a day to 115 million barrels a day within twenty years. The estimated world resources will be depleted within 41 to 60 years. Meanwhile, the monstrous industrial machines of countries like China and India are scrambling for more and more oil to fuel their explosive economy. Those same countries will have to support 1.3 billion more people in the next twenty years. And the U.S.-- the number-one-ranked global warming polluter (we use 25 percent of all energy resources on the planet)--will be home to about six hundred million people by then.

The next fifty years will tell the story -- and it could have an unhappy ending: a planet that has become too hot - not enough oil to run the world's agriculture and transportation machinery to get food to billions of additional people. Not enough fresh water. Oil, food, and water -- that could be the triple threat that creates the worst disaster that civilization and the world has ever seen.

The collapse of modern civilization is possible, remote, but possible. It has happened before, and it can happen again. Four thousand years ago, when the climate shifted into a dryer, cooler period, the Egyptian civilization suffered prolonged droughts. For decades the Nile did not flood its banks and bring needed nutrients onto the land, crops failed, people starved, and society collapsed. They had to rebuild slowly over the next two centuries.

The people of Easter Island also experienced the decline of their civilization after they destroyed all the palm trees on their island. Without roots to hold the topsoil, the rains washed it away, ending their ability to grow enough food. Their population declined steadily from more than fifteen thousand to 111 people. They probably thought they were helping the economy by cutting down the trees when in reality the seriousness of this mistake disrupted and shortened their lives and forced some of them to resort to cannibalism. Similarly, the Vikings were probably driven from North America and Greenland by the extreme cold and a short growing season caused by the mini-ice age of 1350 to 1700. Napoleon came to power because of the French Revolution, which was caused by the shortage of food during this mini-ice age, thus changing the map of Europe. Climate affects everything and everyone.

This can't happen again, you say? Don't be so sure. We cannot allow the comforts and stability of our modern society to lull us into a complacent feeling of security when we're facing a potentially devastating climate change. Global warming is a disruption of the environment produced by our own industrial expansion and pollution, and it really could bring our civilization to its knees. Rebuilding after an environmental Armageddon may be an option, as the Egyptians had, but they had to suffer through two centuries of widespread upheaval, civil wars, disease, infanticide, and cannibalism. In an age of nuclear weapons, rebuilding may not be as possible for our civilization as it was for theirs.

Our window of opportunity for correcting our course before it's too late is very narrow--from ten years to a few decades at the most. We must stop "cutting down our palm trees" (i.e., increasing CO_2 emissions and organize a global grassroots effort to make the necessary changes). We must find alternate energy sources sooner rather than later and hold globalization in check until the scientists complete their studies and we're on solid environmental ground again.

Will our civilization meet its own demise before we manage to take appropriate action? That answer will come only in time.

The debate's over: Globe *is* warming

Politicians, corporations and religious groups differ mainly on how to fix the problem

By Dan Vergano
USA TODAY

Don't look now, but the ground has shifted on global warming. After decades of debate over whether the planet is heating and, if so, whose fault it is, divergent groups are joining hands with little fanfare to deal with a problem they say people can no longer avoid.

Cover story

General Electric is the latest big corporate convert; politicians at the state and national level are looking for solutions; and religious groups are taking philosophical and financial stands to slow the progression of climate change.

They agree that the problem is real. A recent study led by James Hansen of the NASA Goddard Institute for Space Studies confirms that, because of carbon dioxide emissions and other greenhouse gases, Earth is trapping more energy from the sun than it is re-

Please see COVER STORY next page ▸

A warming world

This simulation compares air temperatures near Earth's surface during the last 20 years of the 20th century with projections of temperatures during the last 20 years of the 21st century. The greatest warming occurs in the Arctic and Antarctica.

Projected temperature increase from 2000 to 2100

0 1.8 3.6 5.4 7.2 °Fahrenheit

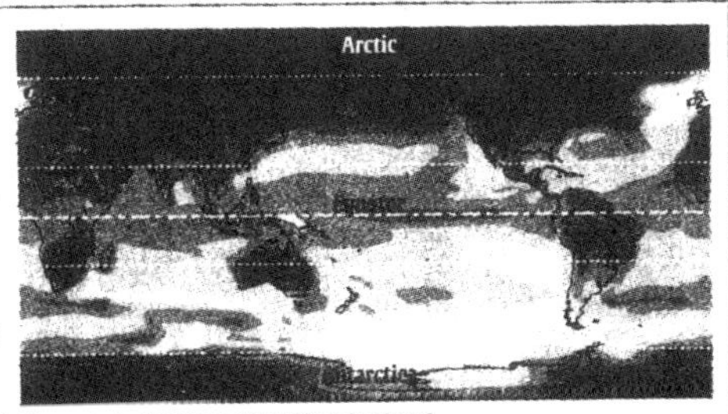

Note: The Community Climate System Model map is the product of computer simulations developed at the National Center for Atmospheric Research with input from university and federal climate scientists.

Source: National Center for Atmospheric Research

USA TODAY

CHAPTER FOUR

Life at 110 Degrees

In the Australian desert, temperatures can reach 140 degrees Fahrenheit. One way the people there get relief is by going underground. Homes, bars, and even hotels are built ten feet below the surface, creating an environment that keeps temperatures, in most cases, at a comfortable eighty to ninety degrees. During the day all human activities are, for the most part, conducted underground. The temperature at night may drop only to a hundred degrees. People come out at night and do what they would ordinarily do during the day. Entire towns, especially in the northern desert known for its opal mining, live this way.

If the temperature of the Earth warms up seven to ten degrees and perhaps abruptly spikes to fifteen degrees or more than the average, the changes in our lifestyles will be dramatic, indeed. More activities will be done in the cooler evening hours or late at night--or even underground.

In Las Vegas, during the Summer months, when temperatures can reach 115 degrees or more, most local residents spend their daytime hours indoors. Except for the die-hard tourists, many visitors swear they will never again visit Las Vegas in July or August. It is simply too hot--hot enough to fry an egg on the sidewalk--but of course it's a dry heat, as they say. Dry or not, it's hot, and high temperatures can be deadly. Three or four children die every year in Las Vegas when parents leave them in a car for just a short span of fifteen to thirty minutes. But by then it's too late, because the temperature can rise quickly above 120 degrees. The same thing happens all too often with pets left in an overheated car.

In 1995, the Chicago heat wave killed 465 people, and in the summer of 2003, about thirty-five thousand people, many of them elderly, died in Europe because of the unusually stifling heat wave that enveloped that region. More recently, the heat killed twenty-two people in the Midwest during one week in July 2006, plus eighty-three others in California that same month. When the temperature climbs to high levels, the very young, the elderly, and the sick are at greatest risk. Rarely do the countries in Europe experience temperatures of more than a hundred degrees, so air conditioning is not commonplace in many buildings, homes, and automobiles. The heat wave of 2003 caught many people off guard, and they were ill-prepared to handle the dehydration and subsequent exhaustion. Even at night the temperature hovered near the hundred-degree mark, allowing no relief. People died so quickly that the backhoes digging the graves could not dig them fast enough to bury the dead.

If the temperature of the world increased ten degrees, you could surpass the number of those fatalities in France by hundreds of thousands, perhaps millions, because people in many countries exist without air conditioning and have no way to escape the heat. They just won't survive. Even in economically advanced countries like the United States, many elderly and sick people would be affected, depending on the city in which they live. Los Angeles, for example, has poor air quality. While the residents usually enjoy a mild year-round climate, on unusually hot days the air quality becomes dangerous for just about everyone. The heat mixed with the smog creates a condition similar to emphysema, where the lungs cannot get enough air to supply sufficient oxygen to the bloodstream, causing hypertension. As many as three thousand to five thousand people die each year from heat-related symptoms and air pollution in Southern California. Global warming, even at these lower levels, currently causes an estimated one hundred and sixty thousand deaths a year worldwide.

The increase of carbon monoxide in the atmosphere is another critical problem that has accelerated over the past few years. In the early 1990s, carbon monoxide emissions were measured at five thousand particles per cubic inch; by 1998, that number had risen to 5,850 per cubic inch, and it has increased rapidly since then. The Environmental Protection Agency (EPA) released a statement saying that it does not consider carbon monoxide to be a pollutant and that they have no jurisdiction over supposed greenhouse gases. This conclusion allows the automobile industry and aging factories to continue operating without refitting with newer equipment that reduces these dangerous emissions.

The bottom line, however, is just this: heat kills. Unless we're prepared to deal with temperatures of 110 degrees and higher, without proper air conditioning, we're destined to lose many more people every year.

Scientists around the world agree that we're heading for the most significant change in temperature in the last thousand years, since the last European mini-ice age, which took place from 1350 to 1700. The warming atmosphere will have a gradually increasing impact on us in the next century. Combine this with the depletion of oil reserves and food shortages caused by heat-related crop failures, a population explosion (with three to nine billion more mouths to feed), and the warming of the oceans by 3.5 to five degrees to the point of a major methane release, and we have the perfect recipe for a global cataclysm of truly biblical proportions--an ecological Armageddon worthy of the most cataclysmic prophecies in the Old Testament and yet, despite all the warning signs, we continue to sail merrily along on our fossil fuel ship into uncharted atmospheric waters that no one in all of recorded history has explored.

Is this our best definition of progress? They say you can't stop it, and they may be correct, but we can change its direction, slow it down, and let the scientists do their analyses and find the best option for us to arrive at a different place before the deadly consequences are unavoidable. Until we find a way to do that, we are continuing to perform a perilously grand experiment and betting the farm with our children's future at stake--and the laboratory is called Earth.

A Bleak Prognosis

Perhaps you think that I have been overstating my case. If so, I will remind you that all but a few of those in the thinning ranks of doubting Thomases now believe, and they see a poor prognosis for the future health of Mother Earth. The environmental thermostat has indeed been cranked up, and we will remain on a trend of an increasingly warmer hot seat throughout this century until we see the world's average temperature increase by as much as 5.4 to 10.3 degrees F. Winters that are 10 degrees warmer might be a pleasant prospect, but what about an additional ten degrees in the Summer, especially in the warmer regions? That could be devastating. A 20 percent increase in skin cancer is hardly worth mentioning compared to the vast disruption of harvesting cycles, serious water shortages, and droughts that are worse than anything we've ever experienced. Such conditions would play havoc with the global economy. Rising temperatures would also change the saline content of the oceans and affect all fish and mammals. Food supplies would dwindle, and the prices for food, electricity, and water would skyrocket. Lakes, streams, and forests would dry up and cease to exist. Serious regional wars would flare up over dwindling oil, water, and food supplies--and this time warfare could include the horrors of nuclear weapons.

In light of that prospect, why don't we go ahead and set the stage for the perfect disaster right now? Let's welcome Iran along with North Korea into the nuclear club, and they can encourage another ten to fifteen nations to join us by the end of this century. That would ratchet up the stakes quite a bit. As the bickering over dwindling natural resources and oil supplies increases, the risk of a nuclear war goes up exponentially.

We are heading into a very interesting century, and you have a front-row seat to watch the unfolding consequences of 57 billion tons of carbon dioxide (CO_2) flooding into our atmosphere each year and steadily increasing temperatures, the consequences of which no one can entirely predict. But since we cannot put oil and gas back in the ground or unmelt a glacier, we may be able to do very little once we tip the scales too far and pass the point of no return.

Although no humans may be around to watch the show, the Earth will eventually find ways to readjust itself and return to equilibrium. Some living things will thrive in this new world, but most others will be gone forever. The average life span of a species is one million years; humans have been around for about five million years, so we could give ourselves the benefit of the doubt and assume that some people would survive a sudden and dramatic climatic shift. The ultimate cost, however, would be an enormous loss of life, if enough of civilization survives. Some people may even benefit from the new order of things. When we slaughtered fifty million buffalo in the American west, the hunters, skinners, tanners, workers in the apparel industry, the government, and the army all enjoyed the benefits of removing the main food source of the Native Americans. The big losers, of course, were the buffalo and the Indians, who were decimated and starved onto reservations.

If we don't destroy civilization completely, others may find themselves in the winners' circle because of a major climate shift. These include the water and electric companies, food processors and distributors, air pollution control companies, and others. The stock market would also find plenty of ways to capitalize on the upcoming debacle. Many smart-minded people on Wall Street are already thinking about how to make money on the future changes. They know that money can be made on any disaster, including war and famine. Security stocks and taser stun-gun stocks became hot items after 9/11, for instance, and sales of heaters, TV sets, and duct tape spiked in New Orleans after Hurricane Katrina in 2005. The moneymen know they can turn a handsome profit by investing in businesses that are at the forefront of stemming the rising tide of problems caused by a world that has become too hot. All they have to do is figure out which companies will benefit most from a disaster. Making a profit out of some poor soul's misfortune is always the feast of the economic vultures.

The losers will be the insurance companies that have to cover the catastrophic losses. Those companies got just a little taste of things to come with the three back-to-back hurricanes that hit Florida in 2004. Tens of billions of dollars in claims bankrupted more than a few insurers. The 1,700 tornadoes and the fourteen major tropical storms that occurred in 2005, including seven category-three or greater hurricanes and made the insurers tremble. The people who live in poor countries will be among the losers too, because they won't have the money to buy the food and water and technology they need to survive; they will simply die by the millions. Many others will be forced to abandon their flooded coastal villages and island homes.

The legacy of the twenty-first century will be a hotter, thirstier, and more dangerous environment where every passing year will witness the loss of more forests, plants, animals, and natural habitats. People will be left with their DVDs and plasma TVs and the images of what it was like to have clean, virgin rivers, natural forests that seemed to go on forever, and seas with healthy coral reefs teeming with life.

Those who survive the ecological fall will have to live with our mistakes and lack of vision. Surely they will remember us as the most irresponsible, self-indulgent generation that grossly miscalculated the impact of interfering with our ecological underpinnings, mindlessly allowing our planet to lose its fragile natural balance and causing the largest mass extinction since a meteor set into motion the domino effect of an ecological Armageddon that all but annihilated life on Earth sixty-five million years ago and came close to decimating it completely 248 million years ago.

The Bible says that our world will be destroyed by fire. Before now, imagining such an event seemed impossible. But if we stay on our present course, what seemed unimaginable could become all too real. The fires of 2003 in California offer us a glimpse of that horror. Those infernos raged across vast expanses of land. If the firefighters had not contained the blazes when they did, many more lives and homes would have been lost. That could easily happen next time. Every Summer of the first five years of this century, the forest fires have gotten worse because of the hotter and drier conditions. Add the threat of nuclear war, and the impossible to imagine becomes a possibility, no matter how remote.

Our fate boils down to a simple choice. Do we want to live in harmony with nature or do we want to overexpand and ultimately destroy it? In the futuristic movie *Soylent Green*, Edward G. Robinson goes to a suicide center, where he is shown images of green hills covered with flowers, flowing streams, and wildlife. All he can say is "Beautiful . . . it's so beautiful!" He had known the world only as a concrete jungle his whole life.

If that's our future, who would want it--or want to leave that legacy to future generations? Of course that assumes that we will have future generations that are still here to inherit it.

CHAPTER FIVE

The Perils of Overpopulation

As if we don't have enough to worry about already, the increasing stress of a burgeoning population is adding to our global woes. By 2020, the Earth will be home to at least 7.2 billion people, and the population is expected to swell to 9.5 billion by the end of this century. Having so many people just increases the tribulations of life in the twenty-first century, when the combined forces of global warming and overpopulation will send us sliding toward the abyss.

Contributing to this flood of humanity are several significant trends. First, people are living longer. The average age today is 82 for women and 74.6 for men, but life expectancy is sure to increase further, because genetic, biotechnological, and nutritional advances will continue to break down the barriers of aging. By 2020, we may feasibly predict that geneticists will have cracked the DNA codes that control aging and will also be able to build replacement organs from stem cells. Thus a human lifespan of one hundred years or more seems plausible within this century.

Second, modern societies will not be able to sustain their rising costs of health care. Right now two of the wealthiest people in the world are pledging billions of dollars to combat diseases like malaria, AIDS, and tuberculosis, to feed the hungry, and to reduce infant mortality rates. Although these efforts are driven by humanitarian zeal and the best of intentions, the results will have some harmful effects. Improved health conditions will save many lives, but that will also add to the problems of overpopulation by pushing us toward that nine billion head count faster than anticipated. Having more people alive and reproducing will increase CO_2 emissions and global warming, thereby imperiling the lives of the very people for whom the health benefits were intended. A few of the billions that they're so generously contributing to healthcare would be better spent on solving the dilemmas of global warming. If they did that, then many of the people they want to save now will still be around to enjoy their lives later in this century.

Another trend: as people live longer, the definition of old age will change. No more will it always be a time when you aren't able to do what you once did. Having an active and vital life to age 85 and beyond is definitely within our reach. With enriched foods, megavitamins, herbal remedies, miracle drugs, and cosmetic surgery for just about anything you can envision, and new pharmaceuticals like Viagra available to everyone, life at eighty-five will be as satisfying and stimulating as it is at age 55 today.

On the down side of this trend will come a number of profound effects on human society, including the problem of dealing with health care costs associated with a graying population, especially since the work force will increasingly become older. Increased pension costs and a reduction in the relative size of the working population will most certainly strain the social fabric. Living in an increasingly complex, fast-paced, high-tech, and computer-driven world is cranking up the level of anxiety. High-density population centers, with their congested traffic, pollution, and crime, will continue to fray our nerves and try our patience. Moreover, our feeling of security may be reduced because of terrorist activities or the likelihood of another major war.

The most immediate challenge, though, will be feeding these additional billions of people. We have only so much land available for cultivation, and crop yields can be pushed only so far. The bounty of the sea isn't nearly as generous as it once was. Because of overfishing, we are already at the point where thirteen of the seventeen major commercial fisheries are on the verge of collapse, meaning that they have been fished to the point where they can no longer reproduce enough to replace what we harvest. Each year a number of species are almost fished out. We have also created dead zones in the oceans from pollution that stretch ten to twenty miles out to sea, where virtually nothing can survive but microscopic scavengers.

Food shortages take a heavy toll on humankind today. Millions of children go to bed hungry and hopeless every night. Thirty thousand children less than five years of age die each day from starvation and disease. Those numbers could increase tragically in the years just ahead.

As the world gets hotter and we add three billion new people to the planet in the coming decades, all these people will also need plenty of fresh water, in addition to food, to survive. Where will it come from? Worldwide, we have lowered our underground aquifers by about one-third, and many of them are no longer able to replenish what has been removed. The lack of potable water has become a crisis in many countries; within twenty years it will undoubtedly be a major problem in the U.S., as well.

With the threat of inadequate supplies of food, water, and other natural resources looming on the horizon, instability rather than stability will be the most likely scenario for the twenty-first century. Do we really need to increase the population of human beings by 32 percent--and all the misery that goes along with that--before we find answers to these pressing problems?

Truth and Consequences

We are adding eighty million new people each and every year to a world that already sees thousands of children die every day from hunger, disease, or the lack of clean, drinkable water. In the past, civilizations were relatively isolated, so the consequences of overpopulation, most notably crossing the sustainable threshold of the regional ecosystem, had mostly a local impact. Today we can cross these ecological boundaries in a very short time. The effects are felt quickly, and often they have a national impact or even a worldwide effect.

For example, the collapse of the cod fishing industry affected all the other fisheries of the world. Nevertheless, fishing trawlers with five-mile-wide nets that scoop up everything in their path continue to wreak havoc on sea life. After decimating one species, we're forced to find new ones to exploit, which then puts too much pressure on that population. Ultimately, that will leave the great oceans of the world all but empty. Species such as tuna, swordfish, orange roughy, sharks, and even the Patagonia toothfish are all in a steep decline. Clearly, this is a case where the domino effect of overfishing will ripple across all the oceans of the world.

Our use of freshwater resources offers us another good example of how one part of the ecosystem impacts another. We have polluted many rivers, lakes, and streams, and we have learned that we cannot contaminate one place without putting more pressure on another. Our seemingly insatiable thirst is rapidly approaching an unsustainable limit. Will water from Canada become the next major export? Like OPEC, will Canada be shipping barrels of water to the U.S. and other arid nations? If so, will we have enough money to buy it? A quart of water from Canada at your local convenience store costs more than a gallon of gasoline shipped from the Middle East. At least that was the case before gas prices jumped 73 percent in just one year to more than $3 a gallon.

What about air pollution? When polluted air drifts out of Los Angeles and Mexico City, where does it go - into Nevada and Arizona, of course. In effect, we are exporting poisonous air (and polluted water) to other cities and countries.

The two main problems are the rapid increase of population and the globalization of commerce, which adds eighty million new consumers and millions of new automobile owners to the planet each year. In 1950, we had 56 million automobiles on the road; today we have 564 million, and that number goes up by 17 percent annually. Second is the globalization of commerce. The Internet has cast its web across the whole world and disseminated ideas as much as information. Now more people than ever want to live by U.S. or European standards, which means that we're depleting even more resources to provide them with cars, cell phones, and sneakers. Whole ecosystems are being wiped out in direct proportion to this exploding consumerism. Sooner or later, diminishing resources will drive up the cost of goods, fueling inflation, while fewer countries will be able to buy enough oil, food, and fresh water to meet their needs.

Where will it end? Because the destruction of one habitat or ecosystem transfers the stress to another one, eventually the ever-growing demand for basic necessities and more consumer goods will push the planet's ecosystems, their fragile habitats, and their inhabitants, to the breaking point.

Does this just sound like so much gloom and doom? If you think so, remember that scientists estimate that we're losing approximately a hundred species a day. That's seven to nine thousand species by the end of the twenty-first century. That means they are gone forever, which tells us only one thing: our consumption and our destructive habits have begun to take a deadly toll on the other residents of planet Earth--and we just may be next. When scientists say that half the species on the planet could be extinct by the end of this century, we must remember that we could lose at least half of our own species, as well.

Earlier I mentioned the die-off of some species of frogs in Central and South America. If that didn't sound very important to you, then how would you feel about losing a whole class of large mammals--or maybe two-thirds of all mammals in the next hundred years, as some scientists have predicted?

Like the aforementioned amphibians, larger species such as birds, tigers, and polar bears are like litmus paper for testing the environment. If they cannot exist, then perhaps we will not be able to, either. At the turn of the twentieth century, thousands of tigers ranged across vast stretches of land from India to Africa. Today, of the eleven big cats, three became extinct in the twentieth century, the last one in 1980. The Burmese tiger population was exterminated because of overhunting, and fewer than twenty-five hundred Bengal tigers remain in the wild. Most tigers are on the endangered-species list because Asians, especially the Chinese, believe that some of the tiger's body parts will cure everything from high blood pressure to impotence. The tigers of the world are being poached out of existence--ground up and put into medicine bottles to be sold in Hong Kong, Tokyo, and Bangkok.

The other problem facing the last of the free-roaming tigers is that the human population of India, Africa, and America is expected to double in the next twenty years. The resulting human need for more land could spell the end for the last of the wild Bengals, as well as Siberian tigers, leopards, cheetahs, mountain lions, and the other great cats that still roam freely. A small number of these cats have found refuge in the few scattered game preserves, zoos, and protected farms, thanks to the work of some dedicated individuals here and there. With luck, their sacrifices will make the difference between absolute extinction and the preservation of a few hundred or a few thousand of an endangered species.

The twentieth and twenty-first centuries will be remembered as a period that devastated and drove to extinction more species than any other era in human history. Hundreds of species have already been exterminated by hunting, overharvesting, and by the overdevelopment and destruction of their habitats. Thousands of other species--some still undiscovered--have been or will be lost because of the destruction of the rainforests. We may never know how many interesting creatures and medical discoveries have been burned up, plowed under, or drowned by the damming of rivers, all to make way for more development.

All these changes will inevitably cause great friction when the relatively few people who consume most of the goods and resources and the vast majority who have little or nothing start to have even less and less. Predictably, one day the multitudes of the have-nots, with nothing left to lose, will rise up to take from the haves. Then, for a time, humankind will have a level playing field again, because war, famine, and disease are, after all, the great equalizers. How can that seem so far off when we're already fighting over oil, water, land, fishing rights, and, of course, religion? Wars over resources and land are as old as civilization itself. We can expect new and even bloodier disputes to arise as we compete for diminishing resources.

Political differences continue to add fuel to the fire, as well, in the Middle East. This is yet another case of the haves versus the have-nots. With growing religious discord between Christians and the Muslim world, we have ideal conditions for a long and bloody conflict. Imagine if all the nations that are at odds with one another currently had nuclear capabilities.

While we fight among ourselves, many other life forms will simply expire as the teeming human population balloons to 9.5 billion in this century.

Are we Making any Progress?

To answer that question, I will reply with a qualified yes, even though the inevitable expansion of the human race is working against us. We have begun to move i the right direction but, unfortunately, we're moving much too slowly and we may already be fifteen to twenty years too late. By now, we should have figured out just how much economic and population growth this planet can sustain. The ideal time to make such determinations, though, was twenty years ago, at the very least. That's when we passed the point where we could have stabilized the population, where each person on the planet could have possibly had the same access to food and clean water and perhaps someday even the same economic benefits of, say, a Swiss citizen. Since we failed to meet that deadline, what price do we have to pay now? War, starvation, disease, economic collapse – quite possibly since these events constitute the traditional ways of culling the herd.

I think the solutions to our problems are within our grasp, but they may be "a bridge too far" if we cannot marshal the political will very soon. During World War II in Europe, the Allies wanted to capture three key bridges at the same time. Seizing and holding these bridges was necessary to allow the Allies (the U.S. and British forces) to cross into Germany. Despite the heroic efforts of many soldiers and a great loss of life, they were able to capture only two of those three bridges. The third one was simply a bridge too far. We have millions of caring, selfless people in this world and thousands of right-thinking organizations that are working hard to make things better. In the face of a relentlessly expanding populace, globalization, and adding three billion more fossil fuel-burning vehicles to the world's highways, their worthy goals may be a bridge too far, and all their efforts may be in vain.

Most people want to leave their progeny behind in a better world – a legacy to be proud of – but what are we leaving for them? I doubt they will think of any of us as "the greatest generation." More likely, they will see us as the greedy and self-serving who live for today. It's all-about-me generations who plundered Earth's treasury and squandered their natural treasury. Future generations won't understand our lack of foresight, our avarice, our gluttonous consumption of the world's natural resources, and the senseless pursuit of economic expansion at all costs. They will see the twentieth century as the turning point and the twenty-first century as the point of no return. They will wonder why we had no charismatic leaders to inspire and mobilize the world's scientific, political, and religious communities to apply the brakes and prevent this runaway environmental train from flying off thc rails.

They will wonder why we didn't care enough to take action before the opportunity to do so was lost forever, when the cost of waiting had such a horrendous price for all of the inhabitants of planet Earth.

CHAPTER SIX

The Impact on the Global Economy

All the changes in the Earth's climate and in society will have a considerable influence on the global economy. Some farsighted people in the business world know this already and are preparing for the hard times ahead. A number of investors and financial institutions have started reshuffling their investment portfolios to take advantage of the forces of global warming and diminishing oil reserves. They know, for instance, that the economy functions much like an ecosystem in that the financial condition of one industry has a trickle-down effect on all the businesses that support it. For example, if General Motors has a bad year, hundreds (and possibly thousands) of businesses, from tire and glass manufacturers to dealerships and gas stations, will all suffer. If the economy of the United States sneezes, the rest of the world's economies get the flu. At least that's how things worked before the rise of the Euro and China's industrial tsunami and our astronomical federal deficit. The U.S. economy is still a heavy hitter, but we aren't the exclusive driver of the world's economy anymore.

Following is a partial list of businesses that will be affected by global warming:

☺ Insurance: Especially property and casualty, because of payouts for damages from hurricanes, tornadoes, floods, and similar events caused by changing weather patterns. Worldwide, claims will force many insurers out of business.

☺ Agriculture: Because of changes in the growing seasons.

☺ Fisheries: Rising water temperatures will alter breeding patterns, which in turn will lead to overfishing and the extinction of many species.

- ☺ Oil and gas companies: Will feel increased pressure when the public becomes more aware of links between industry and the global warming process.

- ☺ Coal and other mining companies: Will continue to operate with high levels of pollution and to attempt to discredit the scientific community so they can keep polluting without additional government and EPA restrictions.

- ☺ Power plants and manufacturing plants: Most affected will be those that use coal as their primary fuel source. like the petroleum and mining industries, they're also doing their best to delay costly refitting that would provide cleaner emissions standards.

- ☺ Logging: As temperatures rise, trees become more brittle, and their growth slows. Also, widespread forest fires will leave millions of acres devoid of plant life and animals.

- ☺ Aluminum: Increased restrictions on smelters and their emissions will probably force older plants to close.

- ☺ Chemical plants: Tighter restrictions could cause the relocation of these plants to countries not participating in international environmental agreements.

- ☺ Steel mills: The older and dirtier mills in the Pittsburgh and Chicago areas that rely on coal could force sharp price increases.

- ☺ The paper industry: Independent mills depend on buying fuel to produce their products.

☺ Cement manufacturing: Added costs would make competing with foreign companies more difficult.

Consumers will have to pay more for the following:

☺ Gasoline--$2.50 to $3.50 per gallon or more (the cost in that range is already in effect). Is $6 to $10 a gallon coming in the near future?

☺ Fuel taxes (sixty cents a gallon or more). A tax of more than $200 per metric ton for using fossil fuel may be imposed. This has been suggested by some think-tank experts who try to divine the political and economic implications of our present situation and the impact that the current drivers of change are having on the whole system.

☺ Food

☺ Electricity

☺ Home heating oil

☺ Transportation (airlines, trucking)

☺ Homes

☺ Home furnishings

☺ Automobiles

☺ Clothing

In addition, in the future everyone may face restrictions on energy use. And, because of global warming, the cost of goods and services will rise right along with the mercury in your thermometer.

Triple Witch

You may have heard the phrase "a triple witching event" in the context of the stock market. It refers to those times when three important economic forces come into play at the same time, which can create vexing problems for the market. If you're an investor, you might be wise to sit on the sidelines during those unusually erratic periods. The triple witching event makes me think of "the perfect storm" that was created by three violent systems that collided off the Northeast coast and brought death and destruction, including the sinking of the sword fishing boat the Andrea Gail with all hands on board, as depicted in the film *The Perfect Storm*. That was an excellent movie, but the storm was a very real and terrible experience for those who lived through it--a triple-witching event that caused great destruction and loss of life

On September 11, 2001, when terrorists attacked the World Trade Center, they were attacking our premier symbol of American business, and when they struck the Pentagon, they were assaulting the symbol of America's military might. They had also planned to strike the White House or Congress--both of which are primary symbols of American democracy. What would have happened if the other United Airlines jet (flight 93) had hit its target instead of crashing into a Pennsylvania field? The final outcome might have been much different. A few incredibly brave passengers probably saved the White House or the alternate target, Congress, and averted exacerbating the disaster. A president without a Congress or a Congress without a president would have caused a great deal more upheaval. If the terrorists had fulfilled their original plan of three attacks -- the World Trade Center, the Pentagon, and the White House -- we would have suffered a triple blow that could have sent us reeling into an economic and political tailspin from which we may never have fully recovered. Even worse, the attacks could have pushed the global economy into a deep and lingering,

1930s - like depression with unforeseen political and economic consequences.

If the White House had been struck, our country may not have had the leadership needed to make the massive infusion of capital and resources necessary to stabilize the financial markets. If we hadn't done that, could we have been irreparably crippled, never to recover our previous economic and political position in the world? Even a great bull can be destroyed with the precise thrust of a thin sword. Similarly, on 9/11, fewer than twenty men armed only with knives, box cutters, and a small bomb nearly brought the most powerful economic and military power in the world to its knees. That didn't quite happen, but they did disrupt our stock market for five days, after which the market didn't stabilize again for months, and commercial air travel and tourism came to a virtual standstill.

The world may never again look at us as an almost invincible economic and military power. The terrorists wielded essentially the same weapons that men used for thousands of years, and yet they came within one airplane strike of crippling a Goliath. Now we know that we aren't as strong and indestructible as we thought, and you can bet that many weaker countries and terrorist organizations around the world took notice. Now they know they don't have to spend billions of dollars to get sophisticated weapons like atomic bombs, ICBMs, aircraft carriers, and F-16s. They just need to strike a crippling blow to a vital national organ. The Chinese know this, and that's why they're planning to defeat the western powers by making what they call a crippling acupuncture-like strike, in the event we ever go to war with them.

Could that really happen to us? History is filled with similar lessons. Only one Trojan horse with a handful of warriors defeated Troy after ten years of siege and many bloody battles had failed. That certainly changed everyone's perception of the power of Troy. Likewise, when Attila the Hun made it all the way to Rome in the twelfth century, the barbarians knew it could be done. So for the next two centuries they chipped away at their objective until Rome was sacked and burned; after that, Rome was no longer the center of power in the world.

More recently, when John Wilkes Booth assassinated President Lincoln, his Confederate co-conspirators also tried to murder the vice-president and the secretary of state at the same time. Booth and his accomplices knew that if they crippled the Lincoln administration. By striking at the three centers of power at the same time, they might replace it with one that was more sympathetic to the Confederacy. Not that much has changed in the last 140 years; we are much more vulnerable than we'd like to believe we are.

Before 9/11, no one dared to think of attacking the most powerful country on Earth. But no more, because perceptions have changed. Now anyone who is willing to pay with his life knows he has a shot at landing a damaging blow on U.S. interests. The barbarians at our gate will view us differently from here on out.

On the plus side, we must remember that history and fate sometimes work together to provide us with the right person for the job at hand. Winston Churchill was that person for Britain during World War II, and so was Franklin D. Roosevelt for the USA. These leaders possessed the qualities needed to pull us together, patch things up, and restore confidence in their nation. The tragic events of September 11 could have easily knocked Wall Street on its ear and sent our economy spinning into oblivion, with all the other stock markets of the world not far behind. Only the prompt, decisive action by President George W. Bush and Congress, to his and their credit, avoided a potentially much larger disaster.

However, we must still keep in mind the one threat that could damage our economy and our way of life so badly that it would change the world as we know it forever. The triple threat of global warming and putting three billion more fossil fuel burning automobiles on the world's highways, increasing the number of coal fire plants by the hundreds, increasing the numbers of people on our already overpopulated planet to the point of having more than 9.5 billion people and tripling the Earth's CO_2 and carbon emissions in the process, triggering a massive methane release. If those three elements converge at some point in the next twenty to fifty years, as they surely will do, if we continue on our present course, then the events of September 11, 2001, will have been a fender-bender compared to the cataclysmic wreck that such a triple-witching eco-bomb would inflict on civilization. A massive methane disassociation spiking up the temperature or the shutdown of the Great Atlantic conveyor belts starting the next ice age would simply be the consequences of the greed, arrogance, and total disregard of the scientific and natural laws of our planet. At 100 dollars a barrel for oil, with one to three trillion barrels still in the ground, energy and big business have at the very least, one hundred trillion dollars worth of reasons to keep this fossil fuel, carbon dioxide, CO_2 emitting, atmosphere polluting train firmly on track through misinformation and buying political influence and on the environment's side we only have those who care.

CHAPTER SEVEN

Creating a Sustainable Planet

What we need to develop is a model for a sustainable planet with enough viable habitats to support all forms of life indefinitely. If we simply continue to expand the population of humans on Earth, the global ecosystem will break down. Then the majority of the people left on the planet will have the quality of life of those now living in Mexico City, Hong Kong, and Rio de Janeiro.

If we continue to overpopulate Earth, then what will we really have accomplished as a species? By plundering our natural resources and wiping out most of the other species, we may find ourselves desperately journeying into space to find a new planet to populate and pollute--and chances are that we won't find one. Just getting to the nearest solar system will be problematic in itself, since it's a thousand light years away.

As the dominant life form on Earth, we have the moral obligation to be good stewards, to take care of what we have, and to make this world the best we possibly can, then pass it along to future generations. We must preserve the oceans, forests, lakes, rivers, and the entire biosphere. We must make positive efforts now to prevent global warming and end the pollution that contaminates the air and destroys the environment. We are bound by an inherent duty to protect all the wildlife for which we are responsible. All the developers, ranchers, farmers, loggers, hunters, commercial fishermen, miners, and oil companies will simply have to accept the environmental limitations of the commons and acknowledge that responsibility. The animals of Earth have the same inherent rights that we do -- the right to life and to pursue their liberties. All creatures had a place in this world long before we humans decided that our interests should override every other species and that annihilating them was our God-given right because, as some people think, we didn't rise to the top of the food chain just to eat salads. But the time has come to discard those egocentric attitudes. Respect for the inalienable right to exist of all living things that inhabit this planet with us is necessary for everyone and everything to survive. We must make the best effort possible to ensure that our natural world never becomes overburdened or damaged beyond the point of repair.

With today's supercomputers and the creative minds in our scientific community, we possess the potential for using new technology to determine how many people can benefit from the natural resources in a given region without exhausting them and without the extermination of the other species there. To do this, we must secure the cooperation of most of the world's religious groups, and we must cross political boundaries to enact laws and develop the teamwork necessary to sustain the equilibrium of our planet without relying on war, famine, and pestilence to do the job for us.

Hopefully, we can get past those horrific ways of controlling our population. Is this possible? Yes, I believe it is. What sane person does not want clean water, clean air, enough good food to eat, and the opportunity to enjoy the natural beauty of our world and the abundance of this extraordinary planet? Nobody really wants more war, famine, and disease. We all want to live in peace and to make a decent living so that we can feed and sustain ourselves and our families. Unfortunately, we are still falling far short of even that modest goal. We have too many starving people who can't survive without dismantling their forests for food and fuel, destroying natural habitats, and killing off endangered species. Blaming these dirt-poor people for contributing to deforestation and the associated loss of wildlife does nothing to solve the problems. Instead, we need a coordinated, comprehensive effort by global organizations to insure that these problems, which affect all of us, can be slowed down, then stopped and eventually reversed.

We need to ask ourselves hard questions like, if you truly believe in God, do you think He created our natural world just so we could pollute it, destroy it, and exterminate its inhabitants? I doubt that He would be happy if we damaged beyond repair in just five thousand years the world that he put us in charge of. Political groups and religious organizations must cooperate on issues that will affect the future of Mankind. What government wants to manage slums festering with pollution, crime, and violence, with not enough resources for its people? Such conditions are a recipe for disease, crime, homicide, and civil war. Finding a workable, sustainable model for every city and every country is in the best interest of all governments. We need to develop master-planned communities for the new "global village," where each family would be required to limit itself to no more than two children.

Every government on Earth must be willing to support a viable, sustainable model, the plan for which must be put on the table for all nations, states, and cities to discuss and negotiate. Governing bodies must create a new Bill of Natural Rights, an environmental and ecological Magna Carta and Constitution for all ecosystems and species in the world rolled into one. This new basis for governing must take into consideration not just the freedom and rights of people but of all species, and consider their inalienable right to exist and to coexist with us on this planet, and must include a high degree of respect for them and for the environment.

To do this we must relearn what the Native Americans and all indigenous people always knew--that we are all connected to the complex web of life, and we must respect the inherent rights of all the inhabitants of our planet, whether human, plant, or animal. All living things must be assured that their environment will not be destroyed and that they will not be hunted, fished, or harvested into extinction.

Doing this would be one of the top ten, if not the single greatest, accomplishment of Mankind--a giant step for our species and an enlightened move in the right direction, away from the threat of extinction for many species that today are just barely hanging on. Such a sustainable model would be a universal movement toward a global utopia, which we could create with our many resources combined with the vision of some of our most inspired leaders. And then, who knows? We may just save our own species in the process.

A model like this could become a reality. God's Garden of Eden was a place where all things were plentiful and where all creatures coexisted in relative harmony--a divine idea. Today we surely must have people with the leadership skills and influence to unite the world and create a global society of caring people and nations. They could create organizations that would hold the respect and reverence for the environment that the Native Americans held for the land and all the life forms on it. This respect for the natural world existed long before "civilized" people came along and took away their land and their ability to exist without money and government subsidies--before religious, military, and business interests and land-hungry governments changed their native way of life forever.

I would like to believe that polar bears will still exist in the wild a hundred years from now, despite global warming and human intrusion. Learning to coexist with all wildlife is a major challenge and a daunting responsibility, but we can do it, and we must do it. This balanced coexistence is already being practiced in Alaska, Canada, and elsewhere. If the grizzly bear and the wolf can survive, then surely that is the best indicator of Mankind's ability to survive as well.

The time has come for us to turn over a new leaf, to change our thinking, and to honor and respect all the creatures, great and small, that share this planet with us. We need to see with new eyes, which would eliminate cruel and wasteful practices like the annual rattlesnake roundups in the American West. The wholesale slaughter of a reptile from which we have little to fear does nothing to help the environment--and does nothing to honor ourselves. During these events, so-called rattlesnake hunters enter the dens of snakes and spray gasoline in their eyes and on their skin. Have you ever felt gasoline on your skin? It burns like a live flame. Then these human predators haul the reptiles from their home and dump them into pits, where they're starved for weeks. Eventually, for the amusement of a paying audience, they chop off their heads. This whole sickening spectacle is presented for the entertainment and profit of a few people, at the great degradation of another species.

Virtually no rattlesnakes remain in many Eastern states, and if we continue such practices, they will be eradicated from most of the Western states, as well. And when they're gone, they will be gone forever--an important and fascinating creature that will never be regarded in awe by any young, nature-loving hiker in the desert. Many will say, "What good is a rattlesnake to us? It may bite someone. Why not kill them?" But with the rattlesnake removed from the food chain, rats and mice will survive in greater numbers than ever before. Every animal has its unique purpose in maintaining the balance of nature. Haven't we already exterminated enough species to haunt us for an eternity?

For the same reasons, many people may never encounter a buffalo, an otter, or a beaver, all of which were quite plentiful in the recent past. Our children may never see a wolf or a grizzly bear in the wild. We wiped out virtually all the large predators and mammals that lived in North America when the white man came here more than four hundred years ago--all but those for which we have established hunting regulations that limit how many are harvested, such as deer and elk. We have robbed future generations of their right to see, to appreciate, and to study many of God's creatures by destroying their habitat and killing them off instead of respecting their right to live and to share our environment. That's why hundreds of species barely hang onto existence and are on the doorstep of oblivion today. Only human beings can claim responsibility for that.

Some people believe that wild creatures don't have the entitlement to be here if their living conflicts with our developments or if they get in the way of what benefits us economically. Such a notion is worse than the acceptance of slavery in America in the nineteenth century. In those days, at least we let the slaves live to work on the plantations. Animals, though, have no guarantee that we will let them survive. They have no right to exist if some farmer, rancher, or rattlesnake roundup promoter feels that animals should not be a part of our environment. Sadly, human predators know that they can simply kill and get rid of these animals. Ultimately, however, human beings will be the ones who will suffer the loss of yet another strand in the fragile web of life that is lost forever until eventually we may destroy enough of the web that we destroy ourselves in the process.

CHAPTER EIGHT

Some Encouraging Trends

Fortunately, not all the news is bad. Some encouraging trends are gaining traction. One of them is the possible slowdown of our runaway population growth. Also, many countries have signed the Kyoto Accord and are trying to reach agreements on reducing the emissions that are causing global warming. Many big corporations and some fourteen states are voluntarily following or bettering the Kyoto limits on emissions. At this writing, however, the United States still hasn't signed this agreement. Without the backing of the U.S., the whole effort could be doomed, because the U.S. is responsible for a quarter of the world's pollution. On the plus side, though, at least some effort has been made to clean up the air in large cities, including Los Angeles, which has shown that Kyoto-style methods can improve air quality. The air is becoming less polluted in some of the major cities in Europe, too.

Ecological concerns are mounting among the populace in a grass-roots groundswell. A number of important environmental and biodiversity programs are now being funded and implemented. So we do have some reason to be more optimistic about mobilizing the world community to solve our problems before it's too late. Although half of the people in the world live on a dollar a day or less, and ten million of them die of starvation every year, at least we're working on these problems now. A long road still stretches ahead, but we seem to be taking the right steps, at last. We can feel encouraged to see so many people and organizations that care and are trying to do something about our threatening environmental problems.

I know that we tend to look at the glass as being half empty instead of half full, because the doomsday scenario is an easily accepted mindset. The world's environmental problems are very complicated and they may seem overwhelming, but we need to examine all the facts, data, and statistics closely to get a clear picture. The solution lies in the caring and in the protection that we give to the Earth--care for its wildlife, care for its people, and the concern we bring for succeeding generations. Obviously, our job is huge and daunting. A great deal of cooperation will be required to get a sharp and comprehensive picture of our problems and find ways to deal with them appropriately.

We must call upon our best scientific minds to provide us with the most lucid and accurate information possible, supported by computer analyses that will tell us how to prioritize our efforts in order to preserve the environment of our planet. People must be educated continually about conserving the natural ecosystem that gives life to all species.

Although we can point to some success stories, we need more of them. Today, approximately 7 percent of all species is about to disappear. In thirty years that figure will rise to 17 percent. As many as a thousand to ten thousand species will become extinct if we don't change the way we do things. Many of these creatures are the large mammals. Being the cause of these extinctions is not a pleasant legacy for us to contemplate leaving to our grandchildren. We simply must find ways to eliminate or reduce dramatically our deleterious effects on the planet and its ecosystems. We can start by stabilizing population growth and eliminating fossil fuel-burning cars and coal burning and coal fire electric plants and we can do this in five to ten years. Adding another three billion fossil fuel-burning automobiles to the roads of the world in the next twenty years, even if they all get 50 to 60 mpg, like the Chinese auto industry is shooting for, is simply committing ecocide by CO_2 and methane gas.

Certainly we have many reasons for concern, but worrying about the future won't help us to accomplish the hard tasks of insuring our future. What can we, as individual citizens of the world, do to help? How about writing or calling members of our Congress and Senate to let them know how we feel about these issues? We can also vote for and/or contribute to the election campaigns of candidates who are working toward solutions and put a shining light on any politician taking money or voting on any measures beneficial to big energy interests. Any positive action is a step forward. Abiding by the slogan "Think globally, act locally," you can start right in your own community. What piece of the environment can you help to protect and preserve? Would starting a petition and notifying your local officials about the need to control growth help? Of course it would. Could you help to elect an environmentally conscious mayor, county commissioner, senator, congressman, or president? Certainly. All you have to do is decide on an issue--and then act. You can change the world by starting with yourself. As Michael Jackson sang in his song "Man in the Mirror," "Make the change."

Money: The Root of All Evil? Maybe Not.

The desire to accumulate wealth on the part of individuals, special-interest groups, and nation-states has contributed immeasurably to the world's woes. We also see that money has its good side, too. As nations, groups, and people become wealthy, they typically pay more attention to the conservation of resources, at the same time taking up causes for the poor, for animals' rights, and for the preservation of the natural world. They see actions such as helping people to improve their life, raising economic standards, and conserving precious resources as worthy goals that benefit everyone. Considered in this light, money is good; while overconsumption, unregulated globalization, and the burning of fossil fuels are bad.

Many people and several countries now see the concept of creating a sustainable planet as the noblest goal of all. As stewards of Earth, each of us can demonstrate that we're not only a thinking, creative being but a caring, conscientious, and considerate one, as well. We have the knowledge and the resources to feed, clothe, and educate everyone and to save the planet at the same time. We don't want our heritage to be that of a depleted, barren world that was ruined by global warming produced by rampant globalization, unchecked population growth, and an addiction to fossil fuels. The Model T was introduced in 1907. In the hundred years since then, gasoline-powered vehicles have changed the atmosphere of our planet, destroying its atmosphere and placing us all in great jeopardy. Oil interests and the auto industry conspired to kill the electric car and push the hydrogen cell out fifteen years. Let's make them change that. Hybrids are not enough.

More good news: 70 to 80 percent of our rainforests remains intact. We have the brains, the technology, and the manpower to tackle any problem. Also on the plus side are the many more people today who really care about the environment, people who are better informed, and who are swelling the ranks and joining the battle to save the planet.

We need people like that, although we would be much better off if we had fewer people altogether. Instead of trying to accommodate billions of additional humans in the near term, we should maintain zero population growth (ZPG) for twenty years and give the Earth time to catch its breath. Then we would have a better chance of solving our environmental and social problems and moving on with a working model for a sustainable planet.

Such a plan would also give us time to raise the educational and economic levels of the current population and allow global planners enough time to figure out what kind of world we want to inhabit in the future and how many people the Earth can support at a reasonable standard of living. Will all the world's religious and political leaders want this to happen? Probably not. Their many opposing beliefs and agendas will stand in the way of easy solutions. Of all these diverse groups, few see the big picture, so getting everyone to agree would be a triumph of diplomacy and political will. If we can help all people to understand the situation thoroughly, they would see that everything they hold dear is in jeopardy and that nothing short of our survival is at stake. In that case, everyone might join forces to make changes in order to avert a disaster of the worst kind. We must dig down deep and find the political resolve to create the sustainable model, and then get the rest of the world to sign on.

Can we find any other good news? Yes, we can! But time is not exactly on our side, although we still have enough of it to do something about global warming and the problem of overpopulation. We still have time to make a difference, to take a look around and decide what kind of world we want to bequeath to future generations. The clock is ticking like a giant eco time bomb, but we can still do something to avoid the horror of an apocalyptic environmental collapse--a literal destruction of our civilization and a living hell on Earth. As the temperature rises higher and higher, Venus, our closest sister planet, broils with an average temperature of 834 degrees Fahrenheit beneath an atmosphere that is made up mostly of CO_2 that may have come from the oceans and water that scientists believe may have been on the surface at one time. But that was eons ago. Now Venus is just a barren, lifeless planet. We do not want to tamper any further with the thermostat that governs all life on our planet. If we do so, Earth could become something like Venus, and we will be long gone.

We have many good, qualified scientists who can help us to use the remaining time wisely. Our top NASA scientists believe we have approximately ten years before we could reach the point of no return, the tipping point in atmospheric gases. Can we reduce global warming? Yes. Can we find adequate substitutes for fossil fuels? Yes. Can we halt the population explosion? Yes. Whether we do these things or not and our fate itself, will be determined in the next thirty-five years. The generations born between 1945 and 1980 will either drive the world to the edge of the cliff and perhaps over it or find a safer road for us to travel. We just need to summon enough motivation to sweep political, religious, and philosophical differences aside and find the collective determination to adjust our course. We have done this many times in our long journey as a species, and we can do it again.

Is this good news or bad news? I think it's good, because we now know the dangers that may lie ahead and what we must do to make this world better for the next generations and because the will of the people is the most powerful political force we have, much like the vast methane deposits which, once unleashed, cannot be contained. We as a species have overcome every obstacle in our path since our humble beginnings about five million years ago on the plains of Africa--or in the Garden of Eden, if that's what you believe. No matter to what God we pray or what you believe in, we're all on the same planet, and the impact will be the same for saint as well as sinner, for Muslim, Jew, and Christian alike. Time is slipping away. We have surpassed the average one-million-year lifespan of a species by a long way. I'd like to think that we're smart enough to keep on going.

CHAPTER NINE

What You Can Do Now

History has taught us that we have always waited until it's almost too late to try to solve major problems, and then we usually use our favorite tool--war. Given that we have had only seven days in all of recorded history when we were not at war somewhere on the planet, we must really like warfare as a problem-solving device. But such conflict cannot solve the biggest problem facing humanity now--global warming. Once we pass the point of no return, we won't be able to do much to stop the momentum that will affect billions of lives.

When we look at the big picture, we know that we need to kick our addiction to oil, find peaceful ways to convert from fossil fuel to alternative, non-polluting energy sources, and stop increasing the population for a few decades by having no more than two children per family. Talk about an overwhelming challenge! What can one person possibly do?

First of all, we should remember that almost everything in the world has been changed for either good or bad by one person--one person with patience and perseverance, one person who cared and sacrificed and inspired others to act. Whether change was created by Jesus Christ, Julius Caesar, or Mothers Against Drunk Driving, everything that happened came about because one person wanted to change things and make a difference.

What can one person do? More than you may think. Remember the slogan "Think globally, act locally." You can join an environmental group; organize a fund-raiser, a rally, or a benefit; do some environmental research and talk to people; write letters to the editor of your local newspaper; write a book; arrange an awareness meeting and invite a knowledgeable speaker; get media attention on a local pollution or environmental problem; or organize a peaceful demonstration. The choices are many and varied. Keep this fact in mind: unless one person does something, nothing happens. Action requires a spark, someone to initiate movement, to be a catalyst, to do something that gets the ball rolling. How small a gathering is or how insignificant the effort may seem, matters little. The atomic bomb was developed because of three short letters written to President Roosevelt by Albert Einstein. Jesus started with just one disciple and ended with twelve good men who changed the world. Hitler started talking to a few men in a beer hall in Munich and almost conquered the world. You and I may not be like Jesus or Hitler or Rosa Parks, who refused to give up her seat on a bus and, with that one courageous act, changed a nation; nevertheless, a small idea backed by passion and action can grow and grow. As cultural anthropologist Margaret Mead said, "Never doubt that a small group of thoughtful, committed citizens can change the world; indeed, it's the only thing that ever has."

So much needs to be done. Someone has to make an effort. Sure, we can leave it up to the other guy to do it. But will he? No. The other guy never will. Something new has always been started by just one person, often a timid or peaceful one, who finally just said, "I've had enough of this!" We all know enough about the world we live in to do something. Taking the first step is the hardest part, because that move takes the most courage and initiative and willpower. But without that first step, we will remain stuck where we are now. Hoping and wishing to turn a dream into reality may be fine, but the key ingredient is action; without it, dreams are just wishful thinking. We can dream of a better world, and we can make it happen. John Lennon thought so when he wrote "Imagine."

Yes, we as individuals can change things in a democracy, because change is an expression of the will of the majority of the people. The politicians will listen and act if enough people make enough noise at the polls. You simply must find ways to communicate with more people who are also concerned about our natural world. We really can have the kind of cities, country, government, and natural world we want and deserve, but we must take action, persevere, and will it into reality.

The basic problem is finding a way to make the cause important, more important than the soft-money contributions and lobbying of the large special-interest businesses and oil companies who use their influence to exploit our environment for profit. The politicians need to know that we see what's going on and that we care about the future of the environment and the well-being of our grandchildren. We must let them know that we're willing to protest, vote, and fight for the kind of world we want future generations to inherit. What is more important than our children's inheritance and the legacy that we pass on? I believe that nothing is more important than that.

All we can do, as individual citizens, is try to do our best to change things for the better before it's too late, to do whatever we can to promote an awareness of the facts and to voice our opinions while we still have the time. We have to act now. If not now, when? If not you, then who? People like you will decide the future of our planet and whether it will be nurturing and bountiful or hot, hostile, and perhaps uninhabitable.

Jim Hanson, the top NASA climatologist, gives us no more than ten years before we pass the point of no return. Hanson is one of the top three climatologists in the entire world who thinks we have every reason to be extremely concerned. I think we should listen to him. Don't expect your government to give you the straight stuff; that's not going to happen. It hasn't given you honest information on a lot of things, including the serious consequences of global warming and Iraq, and it isn't going to start now. Our elected officials always have to put a spin on things or cover them up to suit their political agenda and to protect the interests of Big Business.

One Mind, One Response

In times of crisis, a large population can think and feel as one. The entire nation felt a collective numbness, shock, and fear on September 11, 2001, when terrorists attacked the World Trade Center and the Pentagon. If the president and Congress had not acted quickly to calm our fears and to bolster our confidence, we could have easily slipped into a period of nationwide paralysis and economic tailspin, which would have sent the conservative financial institutions running for cover. But the president took positive steps to quell our collective anxieties by appearing on television two or three times a day and telling us that everything would be okay. No matter what his shortcomings may be (and he has many), President George Bush should be given a great deal of credit for leadership in that time of crisis. Without that kind of confidence-building action, our stunned economy could have created a worldwide depression. Fortunately, the willingness of the government to infuse billions of dollars into our banking institutions and to steady the financial markets probably saved us. Without that reassurance, the U.S. could have plunged into a deep recession that would have taken years instead of months to resolve.

The results could have been similar to the stock market crash of 1929 and the depression that followed and lasted through the 1930s. If our financial institutions had panicked on 9/11 and withdrawn capital from the market instead of pumping billions of dollars back into the economy, the consequences would have been disastrous. The federal government provided an immediate $70 billion transfusion into the economy and the banking system, which buoyed the dollar overseas, by doing what it could to get various stock exchanges around the world back on track, and by getting the big banking organizations to buy huge quantities of stocks, bonds, and other securities to prevent a Wall Street nosedive. Lacking such quick, decisive financial action, we could have entered a dismal period of "locust years" -- probably three to ten years of economic hardship created by fear and a lack of faith in our government and our financial institutions. That was President Bush's finest hour, because we all know that if he is good at one thing, it's spending money.

So, fortunately, we dodged that particular bullet. Unfortunately, we went to war in 2003 to remove Saddam Hussein from power in Iraq. By the end of 2006, that war had cost the U.S. more than $480 billion and more than 3,200 American lives. Sadly, those numbers are still climbing.

As a nation, we tend to think as one in times of crisis, just as we did after Pearl Harbor. The government, Wall Street, and economists all monitor the confidence level of consumers, business leaders, and voters regularly; this is one of the major barometers used to gauge our economic health. When President Kennedy was assassinated, the world was in shock for weeks. The optimism and enthusiasm inspired by living in Camelot evaporated. When JFK died, we all felt a profound sense of loss and grief. Part of our national pride, JFK and everything he represented was irretrievably taken from us. Many people experienced the same feelings with the death of Princess Diana, an international treasure and a universal symbol of elegance, charm, and compassion. When we lose significant people like these, our sorrow is collective, and our confidence is diminished.

Why is such a common human reaction so important? Because in the future when a calamity of the magnitude of global warming occurs, the reaction will be a collective one. A response like this must be anticipated and dealt with by wise leaders--strong and effective people who are willing to do whatever it takes to restore our confidence.

Undeniably, a common emotional thread weaves its way through all of us, carrying an invisible current of communication that resonates through all strata of human society. Thus, for good or ill, we arc all wired together. In fact, many believe that everything on the planet is connected at some basic cellular level. The cosmos and all living things are made up of the same ingredients as the rest of the universe. So we really are all the stuff of stars. Everything is connected and entwined, if you go back far enough in time. Therefore, you are likely to have shared emotional reactions to local emergencies and national disasters. Think of some of the problems we're facing already. Recently we have seen gasoline prices increase almost daily from an average of $1.50 to more than $3 per gallon, and the price is expected to continue to climb. (A cost of $6 a gallon is projected for ethanol, and the price of gasoline would be close behind.) How high does that cost have to rise before consumers react en masse? A shortage of gasoline is also likely in the near future. That may create panic and hoarding, which could leave supermarket shelves bare within five to ten days. If the water supply of your city is disrupted by a natural catastrophe or a terrorist attack, would your family have enough drinkable water to last for several weeks?

Since we all share a primal communications network that allows us to receive the collective signals of insecurity, danger, fear, and panic, you can appreciate the need to be prepared. As the old saying goes. "The time to dig the well is when you don't need the water."

Getting Ready

To prepare for the exigencies of an uncertain future, which could include sharp conflicts over dwindling resources, what can you do to take care of yourself and your family? Should you do what the Mormons do and keep a year's worth of dehydrated food and a three-month supply of drinking water on hand? Perhaps. Short of heading for the hills and adopting the radical lifestyle of a strict survivalist, keeping a six-month to one-year supply of food and water on hand isn't a bad idea. That's the best kind of insurance. Talk to your religious leaders and call or write your political representatives in Washington to tell them that you care about these vitally important matters.

Vote for the next president who has more than an ambitious agenda for the environment and not so much for Big Business, Big Oil, and the military. Join or donate money to environmental groups and conservancy organizations. Visit your local animal shelter or humane society and save one of the twenty-four million dogs and cats that are euthanized each year. When you give one of these animals a shot at life, you will find that money really can buy love sometimes.

What else? Plan to have no more than one or two children. Refuse to go along with religious teachings (or any doctrines) that encourage people to have as many children as possible. These practices were embedded in a society of hundreds or even thousands of years ago, in most cases, when the human population exerted little pressure on the environment. Consider the concept of Zero Population Growth carefully, and then do what is best for you, your children, and your grandchildren. Ask yourself: What would they want me to do for the world now to prepare it for the day when they will have to struggle to survive in it? Wouldn't they want us to bequeath to them a healthy biosphere full of abundance for all, instead of a bankrupt, exhausted, overdeveloped world that cannot sustain its population?

You can also resolve to take an individual stand to do your part to preserve the Earth for future generations. You can refuse to take part in the continuing desecration of our planet, knowing that everything we use in the form of electricity and transportation emits carbon. Shoot for being a zero carbon emitting family. Whatever we destroy leaves that much less for our grandchildren to enjoy. Keep in mind that passing the limits of what we need to maintain adequate supplies of clean air, food, water, and energy is not a good survival strategy for our species and that increasing our numbers without establishing a controlled, sustainable global growth plan is a prescription for environmental bankruptcy. Remember that we homo sapiens are the last representative of perhaps eighteen hominid types that have evolved over the last five million years. Seventeen of them are already extinct. Once we're gone, that's it--no more hominids. We will just be another line of intelligent hominids that failed to heed the environmental warning signs, couldn't adapt, and ultimately failed, like the Neanderthals, Cro-Magnon man, and our other hominid cousins that didn't survive.

Instead of feeling indestructible, we, as a species, should remember that life on Earth has suffered fifty-four major extinctions, fifteen of which annihilated more than 30 percent of all species. The one that occurred sixty-five million years ago not only wiped out the dinosaurs, but also every other creature that was larger than a crocodile. Even worse was the extinction of 95 percent of all life on the planet that occurred 248 million years ago. Since hominids branched off the evolutionary tree much more recently, at least seventeen different humanlike species flowered for a time and then withered away. What happened to them? Any number of things, to be sure. Some evidence suggests, for instance, that both the Cro-Magnons and the even more advanced Homo sapiens competed with the Neanderthals and helped to drive them into extinction. Now we--Homo sapiens--are the last in a long line of what we could call the Great Hominid Experiment, the end of an evolutionary line that extends back more than 7 million years.

Now we, the cleverest species of all, have the means to destroy ourselves with carbon emissions and nuclear energy. Nature has little mercy for an evolutionary misstep. Saying "Oops!" just doesn't quite cut it. As the last hominid, we have pushed our luck way too far, and we will be wading into extremely treacherous new waters in the next thirty to fifty years--a time that will decide our fate. We may be the most successful species this world has ever known and the top dog for now, but we must remember that when you're on top, the only way left to go is down.

We can all do something to keep that downward spiral from happening. Each one of us needs to make an effort to help find a viable population and economic growth model that allows other species and all of our ecosystems to survive. Working in concert with scientists, politicians, and religious leaders, we need to find a new way to design and implement a sustainable planct modcl for continucd lifc on Earth. Doing so will givc us our best chance for survival. Our efforts should be peaceful but determined, so that we do not blow our chances and plunge into an ecological abyss along with the countless other species that became extinct.

What can you do? Just do whatever you can. That's all anyone can do.

"Hominids Timeline"

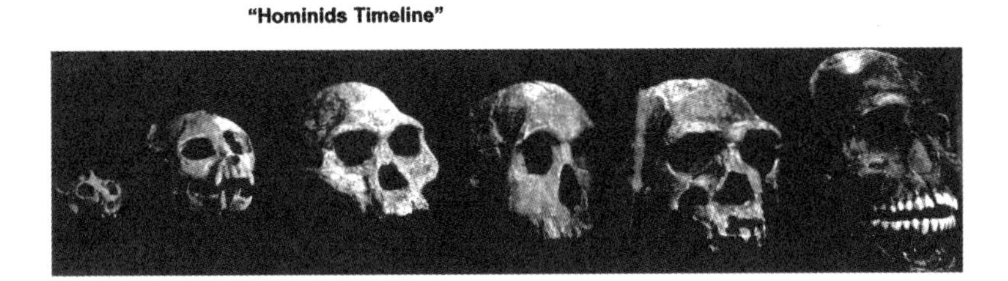

CHAPTER TEN

No Place Like Home

If we could take a giant step back and survey the whole universe, we would no doubt be hard pressed to pinpoint the place that we all call home--planet Earth. All but invisible in the immeasurable vastness of space, our terra firma, our home, is an apparently insignificant speck of matter that whirls around a third-rate sun located in the outback of a galaxy that resides in the boondocks of the universe. When we look at the big picture, this is the humbling truth about our home turf. Considered in that context, our little blue planet doesn't seem like such a big deal.

But let's not be too hasty.

Astronomers have now searched the heavens far and wide to find planets that are similar to Earth--planets that revolve around a sun and possibly include an atmosphere that could support life. Unmanned probes and the Hubble telescope have explored our solar system and peered into the farthest reaches of deep space, giving us glimpses of hundreds of galaxies and discovering scores of new planets.

The only type of planet that scientists have identified so far, however, are the ones that occupy an orbit unlike that of Earth's circular path around the sun. Their elliptical orbits take these planets too close or too far from their sun, making them much too hot or cold to sustain life. No Earthlike planets have been found, even after thirty years of scanning the universe with giant radio telescopes designed to eavesdrop on extraterrestrial transmissions from faraway galaxies. Neither has anyone found any hint of life by using infrared telescopes that can detect even the slightest wobble of light that indicates the presence of a planet in the most remote solar system--the exception to all this being the couple who swear that aliens periodically abduct them for scientific and sexual experimentation.

Such wild claims aside, to this day we have succeeded only in demonstrating that we are unique and that we are, as far as we know now, alone in the universe, a one-of-a-kind experiment in carbon-based life development that requires six key ingredients in exactly the right proportions for life to exist at all. Now we are tampering with this singular experiment by changing the key atmospheric ingredients, thus altering that precise formula that allows us to exist--one planet among millions of others in the universe.

Our equally unique human ego--or God complex, if you prefer--encourages us to believe that we can tinker with or control or conquer anything. We have tamed mighty rivers, leveled mountains and forests, and made ourselves masters of the seas, compelling all of Earth and every living thing to submit to our will. But if we continue to feel that we can dominate everything, riding roughshod over nature in the process, and then we will surely share the fate of the many mighty civilizations that crumbled before us into the dust of the past. We may find ourselves homeless on an inhospitable planet if we are not very careful. Earth is a very special place that can sustain what is perhaps the only life that exists in all the galaxies from here to millions of light years away.

There really is no place like home. If we're smart, we will take better care of it. Every corporate CEO, powerful politician, ordained minister, clergyman, and ordinary common citizen must realize that we are all just a wholly owned subsidiary of the ecological conglomerate called planet Earth--our one and only home.

We must take care of it if it is to take care of us.

Technology Vs. Nature

1912

Titanic

The Marvel of Man's technology hits ice berg on her maiden voyage. Sinks in 2 hours and 40 minutes. As the water rises the passengers move up from deck to deck until there's nowhere left to go. With one grand swirl they sank within singing that sweet melody "Nearer, Nearer My God to thee".

18 life boats - 69% loss of life.

2007

Atmospheric Pollution

Collides with the Environment Despite technological advances it Triggers Global Warming. Temperature increases by 3.5 to 7.4 degrees on average hotter. Massive Methane disassociation causing an additional 5 degree temperature spike. Species that can migrate to higher, cooler ground, do. No life boats. Nowhere to run.

____?____% loss of life.

The Model T Automobile was introduced in 1907

Will Rogers said "In 100 years we'll know if it was good or bad for us."

I believe the verdict is in.

CHAPTER ELEVEN

A Legacy of Misfortune?

Now that we have come this far in our discussion of global warming and its related problems, let's summarize what we know. Here are the most important facts and ideas that we have covered, along with a few more significant points to consider:

☺ The temperature of the Earth and any other planet depends mainly on (1) the amount of sunlight received, (2) the amount of sunlight reflected and returned to space, and (3) the extent to which the atmosphere retains heat.

☺ A natural greenhouse effect keeps the Earth warm enough (with an average temperature of about 58 degrees F.). Greenhouse gases like carbon dioxide, methane, nitrous oxide, and water vapor trap heat and warm the Earth's surface. Exactly how the Earth's climate responds to increased greenhouse gases depends on complex interactions between the atmosphere, the oceans, land, and the polar ice caps--the whole biosphere.

☺ Global temperatures are rising at a steadily increasing pace--a trend that has accelerated since 1950, at about the time that the automobile came into widespread use around the world. Current estimates say that the Earth is now eight-tenths of a degree F. warmer.

☺ Observations made over the last century suggest that the average land surface temperature has risen from .8 to 1.0 degree F. during that time. Since 1979, scientists have generally agreed that a doubling of atmospheric carbon dioxide will increase the Earth's average surface temperature by 5 to 7.4 degrees F.

☺ This calculation does not take into account any spikes incurred due to massive methane releases that could add an immediate 3 to 5 degrees in average temperature.

☺ Each one degree of warming will shift the warmer temperature zones by about a hundred miles northward and to an altitude of about five hundred feet higher. Many plant and animal species, some of which have already retreated to higher ground, will be unable to migrate to a cooler elevation fast enough to survive, and all of them, as well as we humans, will run out of real estate sooner or later. As if climbing the decks of a sinking ship like the Titanic, they will eventually have no more space to move up as the Earth sinks slowly into a warmer and warmer atmosphere. We truly are all passengers on the Titanic as the Irish philosopher pointed out and we must do all we can to avoid passing the tipping point in CO_2 and methane emitters.

☺ Every degree of increased temperature will also reduce the harvest of wheat, corn, and rice by 10 percent. Furthermore, an increase of three degrees C. could threaten seven to 11 percent of North America's plant species.

☺ The northern limit of habitation for many birds is also strongly associated with climate. In addition, scientists estimate that we will lose 1.7 to 2.3 million square miles of cold-water fish habitat by 2060.

☺ The pace at which humanity changes the climate depends a great deal on the rate at which society adds more greenhouse gases to the atmosphere, i.e., with more automobiles that burn fossil fuels and more coal-fire electric plants.

☺ Most of the 54 major mass extinctions documented by scientists occurred because of rapid climate change.

☺ Atmospheric concentrations of greenhouse gases have risen significantly since the beginning of the Industrial Revolution: carbon dioxide by 30 percent, methane by a 100 percent, and nitrous oxide by 15 percent. Many greenhouse gases remain in the atmosphere for a long time, from decades to centuries. Projected CO_2 concentration levels are significantly higher than any estimates for the past 160,000 years.

☺ Until a few decades ago, energy use grew at about the same rate as the Gross Domestic Product, but then that trend spiked upward. Since 1990 we have pumped 1.2 trillion tons of carbon dioxide, a potent greenhouse gas, into the atmosphere each year. This could act as a trigger for a massive methane release.

☺ The amount of methane hydrate in the atmosphere is 1 percent higher than at any time in the last 160,000 years. and has increased 300% in the atmosphere in the last 5 years.

☺ Methane has the very unique quality of letting through sunlight while holding in heat.

☺ Methane is fifty-six times more potent a greenhouse gas than CO_2 or nitrous oxide (the gas generated by burning fossil fuels, including the gasoline in automobiles).

☺ Methane remains in the upper atmosphere for ten to twenty years, while CO_2 stays in the upper atmosphere for 230 years.

Methane deposits are estimated to be as much as a thousand times more abundant than all the oil deposits yet discovered. This represents more than 10 percent of the entire biomass of the Earth.

☺ Methane destabilizes when water pressure falls, such as when periodic ice ages increase the vast frozen areas near the poles and lower the depth of the ocean.

☺ Methane also destabilizes when the ocean temperature increases. Worldwide, the oceans are now one to three degrees warmer at the surface. It takes twenty to thirty years for these heat tracers to reach to ocean bottom and to affect the methane deposits.

☺ As recently as 7,200 years ago, methane destabilization caused two sediment failures (or slope failures) beneath the ocean, the smallest of which caused a mile-high tsunami that swept over the Norwegian coast.

☺ A major release of methane could double the Earth's surface temperature in a very short time. An average increase in Earth temperature of twelve to fifteen degrees F. would prove fatal to most species on the planet.

☺ The ocean temperature is currently rising. The rate of warming has been increasing since 1990.

☺ Temperature increases will raise sea levels by one to three feet in this century. Higher temperatures raise the sea level by melting polar ice fields and the Greenland ice sheet. Sea levels have risen worldwide by approximately fifteen to twenty centimeters (six to eight inches) in the last century. About two to five centimeters (one to two inches) of the increase was caused by the melting terrestrial glaciers. Another two to seven centimeters came from the expansion of ocean water caused by warmer ocean temperatures. If the Greenland ice sheet melts, the sea level will rise by more than twenty feet. If the Antarctic ice melts, add another 200 feet. A disaster like that would change all life or whatever is left of life on Earth forever more.

☺ The North Atlantic is experiencing a forty-year trend of desalinization while the salinity of tropical waters is increasing, according to Dr. Ruth Curry of the Woods Hole Oceanographic Institution at Cape Cod, Massachusetts, and nineteen British and Canadian scientists. They note that the Great North Atlantic conveyor belt, which drives much of the global climate, has slowed by 15 to 20 percent and is being pushed toward a dangerous tipping-point threshold that could shut down the whole belt, precipitating the next ice age in just three or four decades. Any further warming could cause a meltdown of the Greenland ice sheet, which lost 150 feet of thickness in just one recent year alone.

☺ In 2005 a British conference on climate change concluded that a global temperature rise of just three degrees would most likely lead to the melting of Greenland's polar ice. Stephen Schneider of Stanford University has put the odds of a massive Greenland melt at fifty-fifty. In addition, a study described in the journal Science in March, 2006 said that the melting in Greenland is proceeding more rapidly than originally thought and that it alone could trigger a one to three-foot rise in global ocean levels.

☺ An unusually large outflow of fresh water from Greenland and Russia could disrupt the entire North Atlantic oceanic belt, which is essential to our maintaining a stable climate throughout the world.

- ☺ A sea temperature increase of five to eight degrees of surface temperature would take twenty to thirty years to trickle down to the bottom. This could cause disassociation of major methane deposits close to the surface of the continental shelf in Oregon, California, the Gulf of Mexico, Texas, Louisiana, Florida, and other parts of the world, causing an immediate temperature spike.

- ☺ A two to four degree C. increase in ocean bottom water temperature (in the Santa Barbara Basin, for example) would most likely cause the disassociation and de-roofing of gas hydrate deposits there.

- ☺ Kennett, et al, (2000) have suggested a time lag of twenty to thirty years after the surface of the ocean is warmed before the higher temperature reaches the bottom and causes methane release.

- ☺ Fifty-five million years ago, in the late Paleocene period, widespread volcanic eruptions greatly increased carbon dioxide in the atmosphere, thereby warming the ocean depths by four degrees C. and triggering a catastrophic release of methane into the atmosphere, which caused a warming spike that led to the mass extinctions of that time.

- ☺ Two hundred fifty million years ago a meteor slammed into the earth off the Australian coast, setting off a chain of massive volcanic eruptions in Siberia that pumped tremendous amounts of CO_2 into the atmosphere. The additional CO_2 raised the Earth's average temperature by seven to ten degrees, warming the oceans and melting the polar ice caps. The warming of the ocean depths by four degrees C. caused a vast methane hydrate disassociation, sending tremendous amounts of methane gas into the atmosphere and raising global temperatures an additional seven to fifteen degrees F. As a result, 95 percent of all ocean and terrestrial life perished.

☺ The Earth's current population of 6.3 billion people is expected to swell to between 9.5 and 10.5 billion by the year 2050, and the number of cars on the road will increase from eight hundred million to 3.3 billion. Exxon Mobil, the world's largest energy company, says that by 2070, fossil fuels will still make up 84 percent of our energy use even though we are currently consuming two barrels of oil for every single barrel of new oil discovered. We will add 1,100 new coal fire burning energy plants in China, India, and in the U.S. Even a first-year freshman in chemistry can calculate the consequences of that much change in atmospheric gas. We are simply slowly changing our atmosphere into a CO_2 methane dominated atmosphere that existed at the beginnings of our planet 500 million years ago when only bacteria and single cell organisms could exist. And the world was 30% hotter.

Another problem that we could face soon is the increased spread of disease, because the geographic range and life cycles of pathogens and vectors (like mosquitoes) that transmit disease are affected by climate. A warming climate would increase the potential transmission of many vector-borne diseases. Outbreaks of infectious diseases have been associated with specific weather patterns, including:

☺ Malaria: when the weather is hotter and more humid than usual.

☺ Hantavirus: six years of drought followed by a warm, wet spring.

☺ St. Louis encephalitis: conditions that are warmer and

wetter than usual.

All of these potential threats considered, we are compelled to ask, Are these the kind of challenges to survival that we want our children and grandchildren to deal with? Will this planet resemble the world in which we grew up? Will we still have four distinct seasons? Will fresh water still be abundant enough for drinking, raising crops, swimming, and fishing? Will our descendants be able to see and explore forests with wildlife in them? Will they be able to travel down roads that aren't gridlocked with bumper-to-bumper cars or breathe unpolluted air? Will many of our young suffer from chronic asthma or the old die too young from emphysema? Will they inherit a virtual high-tech world where natural resources have been devoured by an ever increasing population whose appetite for global economic expansion knows no bounds?

The Global Future

In 1999, the National Intelligence Council (NIC), part of the CIA's Global Futures Project, convened a prestigious group of government officials and nongovernmental agencies to peer into the future. They did not like what they saw. Their conclusion was presented in the CIA's report "Alternative Global Futures--2000 to 2015." They agreed that we are already too late to have a world population that would allow everyone to have enough food to eat and clean water to drink. We passed that population benchmark, they said, thirty to forty years ago. Furthermore, we will need at least twenty to thirty years to reduce the current number of hungry people in the world by half. This reduction is possible, however, only if global economies and crop yields remain the same. But global warming could rapidly change the growing seasons and reduce harvests.

As limited resources meet unrestricted globalization and economic expansion, we can clearly see that our natural environment is on the losing side of this delicate equation unless a balance can be found. This will be a monumental challenge for humanity, because we are far from the point of equilibrium already. For each person alive today to have the same standard of living as an average U.S. citizen, we would need four planet Earths to supply enough raw materials and natural resources. Put another way, we already have seven times too many people on the planet to allow everyone the opportunity to live the lifestyle of the average European. And we're adding seventy-five thousand new consumers to the planet each day. As of the last census, 6.3 billion people inhabited the Earth. By 2032, the population will grow to nine billion or more on this relatively small planet.

What will happen when more and more people want the same consumer goods that we have in the U.S. when we have fewer and fewer natural resources to supply those goods? With increasing temperatures, dwindling supplies of oil, exhausted aquifers, and less fresh drinking water, and less food, as a result of radical changes in the growing season caused by increased global warming, and an exploding world population, we have a recipe for the perfect disaster--increasing global conflicts combined with the threat of nuclear proliferation.

Considering these future problems is a lot like riding on an airplane that has lost an engine; you know something is wrong, but what can you do besides hang on for the white-knuckle ride?

We're confined to a small planet that speeds along at more than sixteen thousand miles an hour through space with a thin smear of life clinging to its surface. Six essential elements must be available on a planet in exactly the right proportions for life to exist at all. Make a slight change to any of them and life as we know it could not exist on Earth. We would become another Mars, Pluto, or Venus, devoid of life -- just a desolate rock.

As we hurl through space aboard our tiny, blue vessel, headed for an unknown destiny, a potentially catastrophic problem looms ahead of us, warnings of which are impacting us right now, with worse things to come in the relatively near future. After denying the signs of global warming for decades, our government has finally admitted that a problem exists. Now the key questions are: Can something be done before it's too late? Is anyone in our government leading the charge to face the issues and find solutions to the problem?

The truth is that we aren't slowing global economic expansion enough to study the problem properly and to find solutions. In fact, for the last twenty years, the U.S. government has spent a great deal of time, energy, and money creating controversy and rewriting or putting a spin on scientific data in its attempt to deny or try to prove that the problem doesn't even exist at all. President George W. Bush even appointed a man from the petroleum industry, Phillip A. Cooney, as the head of the Department of the Environment, even though he had no previous scientific experience. Nevertheless, he edited or rewrote a number of scientific research reports for the White House, taking definitive scientific analyses and rewriting them to lessen their credibility and encourage controversy. When this situation came to light, Cooney quit his government position and went to work for ExxonMobil.

Of course our leaders knew what was happening. They have known for some time now. The CIA's 1999 global trends reports say that members of a government think tank have told them what is likely to happen if we stay on this course. Nevertheless, the government has chosen to support a fossil-fuel strategy that protects the energy sector. Even though hundreds of scientists have documented and verified the adverse effects of global warming, we're still pursuing economic expansion and globalization and supporting fossil fuels at all costs, making the natural world, which sustains all life, much less important than the stock market, which sustains our economic growth.

Dealing with a few more degrees of heat or a little less water is low on the priority list compared to a trillion-dollar increase in the Gross Domestic Product. So what if a few hundred species die off each year? Isn't that a small price to pay for a robust economy? Mass extinction hardly compares to the importance of another $500 billion for a military that can already put a laser-guided missile into your toilet bowl from two hundred miles away or destroy the Earth a hundred times over with the weapons of mass destruction it already possesses.

Yes, having a healthy economy and a capable military is very important, but such things must be balanced by how we develop our world and what effect that will have on how our children and grandchildren live. What needs to be factored into the economic equation is the idea of a balanced, sustainable world that has enough room and resources for all the other living things that share our planet to survive. The environment and the health of our planet need to take priority over political, religious, and military agendas, not be tenth on the list. Our survival and the quality of our future lives depends on this. In the final analysis we and the natural world will have to pay the price for our unbridled growth.

If global warming is the fundamental problem, the cause of the problem is too many people driving too many vehicles and too many factories and power plants burning coal and other fossil fuels. They're all dumping too much carbon dioxide into the atmosphere. Ironically, fossil fuels are nothing more than the Sun's energy stored in the countless plants and animals that died through the millennia and were slowly converted over hundreds of millions of years into oil, natural gas, and coal. We started to use this stored solar energy in earnest about 1850. Fossil fuels powered the Industrial Revolution and allowed us to clear vast tracts of farmland with sophisticated and powerful new machines. We were able to produce enough food for the population of the world to explode, which, in turn, fueled tremendous economic expansion. We used much of this stored energy in the twentieth century. Today, some experts predict that we may have only enough oil reserves to last through three-quarters of the twenty-first century. 2041 to 2065 is now the best estimate of when we will have depleted the estimated oil resources.

We have begun to use up the last of the easy-to-find and fairly easy-top-reach oil reserves. The days of cheap energy are over. Although this primordial source may seem boundless, it is a finite supply so, when it's gone, it's gone for good. Meanwhile, as we deplete our reserves, oil and all petroleum-based products are becoming more expensive.

The converging problems of reduced supplies of food and fresh water, and uncontrolled population growth could overwhelm the Earth's ability to provide for us and the multitudes of other species that also depend upon a natural and stable environment to survive.

The cumulative effects of all the negative forces in play now may leave us much less time than most scientists have estimated before we reach the "boiling point:" in reality the time that remains could be 30 to 40 percent less than any reasonable prediction advanced to date, because methane releases have not been factored into the equation. If that's the case, then the time frame for our global ecological meltdown should be changed from the period of 2040 to 2070 to the much more imminent period of 2020 to 2035. That gives us less than thirty years – and perhaps no more than twenty, to find ways to avert disaster.

Of course we may have some vast new reservoirs of oil out there somewhere, perhaps under the ocean or beneath the rapidly melting Arctic ice, although no one has found a new elephant oil field in more than twenty years. If such new reservoirs exist and if we can find them and retrieve the oil, we can buy ourselves another twenty to fifty years of pollution and oil addiction. Maybe we can change direction or slow down enough to develop or invest in new energies and avoid the inevitable crunch. With luck we may still have enough time if we focus on the problem now.

If we are sensible, we would develop a sustainable world model, one with a population that is in balance with our natural resources and one that leaves enough unspoiled habitats and ecosystems to support most of the other species on the planet, as well. That is the big challenge. We certainly have the computing power to design a workable model for sustaining the planet. But the question is, can we muster the political will to implement it, are we wise enough to replace dwindling oil resources with new alternatives in time to dodge an ecological bullet and still meet the needs of the seventy-five to eighty million new people each year, will we have enough of a natural world left for the other species to coexist with us?

What kind of world will we leave to future generations? In 2099, when people look back upon the stewardship of our generation in the critical first half of the twenty-first century, will they say they knew, they absolutely knew, but they didn't act in time? Why?

AFTERWARD

This book is a journey that began with my discovery of a CIA report posted on the Internet that was based on discussions with nongovernmental organizations, or NGOs, about the impact of global warming on the world. I was disturbed by what I read, but what I learned later was even more unsettling. Not only did our governmental leaders choose not to act on this information about the potentially devastating consequences of global warming, but they also denied that the problem even existed. Until 2004, the Bush administration vigorously maintained the position that global warming was not being caused by human activity and announced that it stood firmly behind America's fossil fuel policy. Only well into his second term, in the face of overwhelming scientific evidence, did the president admit that global warming was caused by human activity and that America is "addicted to oil." As a result, we missed a very crucial window of opportunity. We could have acted sooner to try to find solutions to the problem.

On June 13, 2005, USA Today announced that the world's scientific community had ended the debate about the phenomenon of global warming. The world is indeed warming, the national newspaper stated, because of man's activities and the burning of fossil fuels. A Time magazine cover story in April, 2006 echoed the same news, declaring flatly that the global warming debate is over--and has been for some time.

The biggest of all the potential problems we face, however, concerns natural methane gas. This potential slayer of civilizations is more lethal and more likely to destroy us and everything else on the planet than anything else. Vast quantities of this gas lie dormant and invisible, waiting for the trigger event that will unleash its destructive force. If that happens, the effect would be similar to putting a plastic bag over the Earth and filling it with carbon dioxide and methane gas. And the probable result would be the virtual annihilation of sixty-five to ninety percent of all living things above ground. This has happened before. Once this gas disassociates, we have no way to put the effervescent genie back into the bottle. Then we will suffer the consequences of allowing ourselves to be misled by our politicians and Big Energy businesses and then pay a terrible price for our addiction to fossil fuels and our love of the automobile.

Even without a massive methane disassociation, increasing the temperature of the planet by five to seven degrees with CO_2 emissions will have its own dire consequences. Such overheating has occurred due to volcanic activity in the past, scientists believe, and it will surely occur again, with a predictable impact on the planet.

These developments do not paint a pretty picture of our future. Dare we push globalization and CO_2 emissions so far as to make such events possible or even probable? Do we really want to be the ones who are responsible for tinkering with and unleashing the destructive natural elements of this planet, or do we want to be the generation that stands up and declares that we're no longer willing to gamble with the future of our species?

No matter who is really at fault for our environmental dilemma, the U.S. will be blamed for its lack of leadership. The Democrats will blame the Republicans and vice versa. Politicians will blame the scientists or Mother Nature, and religious institutions will say it's God's will or the devil's work. Some will say that this is God's way of ridding the planet of everyone but the true believers who will be raptured away while the rest of us are burned up, no matter what their political or religious belief may be. Just like those on the Titanic, we're all on the same boat together, so the Republicans and Democrats, the devout and the non-believers will suffer the same fate.

The aboriginal and native cultures of the world understood their connection to the tree of life and knew that destroying any branch of it diminished their lives and, more importantly, their chances for survival. We can clear-cut their forests and force them into the cities, erasing thousands of years of cultural heritage in the process. But we can't snap our technological fingers and summarily dismiss all the mandates of Mother Nature that indigenous people knew so well.

Before our time, many sophisticated civilizations disintegrated and vanished from history, but in the distant past each of those events was isolated and went largely unnoticed by the rest of the world. Today we are one vast, interconnected global community; what affects one small part of it affects the whole world. If we choose to poison the well in our own backyard, we can no longer assume that we won't harm our neighbors.

The demise of any civilization, for any reason, is never a pleasant, gradual decline or a rapid ruin. Our present civilization, when it falls, will fall just as hard as those in the past, and the process will be no day at the beach. The traits that now make us feel civilized, almost invincible, and blessed by God will be replaced by the raw instinct to survive at any cost. We will no doubt experience mob violence, theft, murder, infanticide, and cannibalism, combined with people turning on the political and religious leaders who led them so confidently to their demise, all of which have been the telltale reminders of most of the vanished civilizations studied by archaeologists. Any breakdown of society in the future probably won't be any different from those of the past, except that we can add death by heat and possibly radiation poisoning to the long list of horrors.

Where our particular path leads us through the triple threat of global warming, diminishing oil reserves, and increasing world population that is obsessed with automobiles may be hard to predict, but surely we will see nations squabbling over the world's dwindling oil resources. We have already launched our first war of the century with Iraq in an effort to control what's left of those reserves. Without Iraq's substantial oil resources, we all know that we would not have resorted to a military solution there, because we would have gained no underlying benefit by using what used to be called "gunboat diplomacy."

By the year 2040, with the climate rapidly changing, which in turn will affect the growing seasons and fresh water levels, many people will be displaced, and the world will be hard pressed to provide the basics of life for another three billion people. These developments will inevitably lead to more conflict and more wars, even ones in which nuclear weapons are used again. Our military is planning on that, so we should plan on it, too.

The twentieth century saw the Great Depression in the 1930s, followed by Germany's and Japan's joining forces to conquer and divide up the world and all its resources. In Adolph Hitler's unpublished second book, which was discovered recently and authenticated, he described in detail how, after the conquest of Europe, Germany and Japan would jointly take on the United States. If Japan had been able to postpone or stop Hitler's invasion of Russia (which Hitler did to gain control of the oil fields), they may have succeeded in winning a combined war against the Soviet Union and then attacked the United States. Not likely, some would say, but Japan thought that strategy was quite plausible. In the end, those plans were pre-empted by America's joining the war, Russia being much harder to beat than Hitler anticipated. Fearing a stronger Germany should it succeed, Albert Einstein encouraged President Roosevelt to develop the atomic bomb. If Germany hadn't invaded Russia for its oil, it is feasible that we could have ended up facing a combined force, a stronger Germany and Japan attacking the U.S. on its own soil – that was the plan. This could have happened. It was a very real possibility back in 1942. Who knows what the outcome would have been.

As bad as the threat of the Second World War was for civilization, there's an even greater one today. Now we face an even more destructive foe, one that could conceivably cause the extermination of our entire species. The new enemy is a potent combination of the ignorance of nature's forces, arrogance, ambition, and greed, plus a healthy overconfidence in what technology can do. Put another way, "We have met the enemy, and he is us." As a species, we have not only survived but prevailed through every setback and adversity, but now we find ourselves at the most fateful crossroads ever. Will we make the right choices? If we don't, we may write the final chapter to the story of Homo sapiens and the last hominid on planet Earth.

Unless we're willing to limit our population growth and to reduce the proliferation of gasoline-powered vehicles and kick our addiction to fossil fuels, then anyone born from 1990 through the rest of the twenty-first century will have a fair chance of occupying a front-row seat to potentially the most cataclysmic event in more than 248 million years--the ecological meltdown of the natural world. Then all we will be able to do is embrace the horror. Once it starts, we may be powerless to stop it.

The increasing heat will kill all the plants. Then all the plant-eating animals will die -- cows, sheep, birds, etc. -- followed by the predators and then the humans. Yes, some people may survive, especially if they can tolerate living deep below ground, high in the mountains, or in a closed or controlled environment. The reptiles will go last, when no more plants, insects, or birds remain for them to consume. This would be a spectacle unmatched in Mankind's long and colorful history--an unparalleled tragedy in which we may witness the demise of a species that was successful enough to subdue, then overpopulate the planet, invent the internal combustion engine and the computer, travel to distant planets, and maneuver into the position to destroy itself and virtually every living creature on the planet in less than 4.5 million years after its introduction into the ecosystem. We will probably be the first species, however, not only to be aware of our potential demise by tampering with the ecological mechanisms that control the Earth's environment, but actually be able to look out the window of our cars--while we're stalled in a bumper-to-bumper traffic jam--and plainly see the cause of it through pollution, overpopulation, and the burning of fossil fuels.

Meddling with the natural systems that make this planet habitable is certainly not a very good plan for long-term survival. By altering one of the basic ingredients for life by only one or two degrees of variation, we will alter the atmosphere of Earth -- and most likely not for the better. Will the politicians sit on the sidelines, debating for another decade? How can they not realize that the time has come to slow down and let the scientists do their job, to persuade emerging nations to put their globalization plans on hold, or at the very least slow down. Protecting the natural world is job #1 if we are to survive. .

Time flies and time is running out. By 2012 we should have all the knowledge we need to act and change course. If the world again experiences a massive disassociation of methane sometime between 2020 and 2060, that singular, defining event will threaten our very survival, we must be aware and be prepared. By saying this, I'm not just trying to inspire fear or anxiety about what may occur, but I am saying that we still have the time and the options to let the scientists study the problem and then take their advice on how to proceed safely into the future. Think of them as the doctors for the planet and if we will follow their advice, well sustain a healthy planet; otherwise, we will listen to the politicians and keep on burning fossil fuels. Until we trigger a massive methane release that once started can't be stopped, we really shouldn't get Mother Nature too riled up if we can help it. All the same, though, we should prepare for the worst while we hope for the best.

In the great abyss of time, we do not have a limitless amount. It is, by its nature, neither boundless, infinite, nor endless. All of us, from the very beginning of Man up until this very instant, have survived by using all of our senses to warn us of impending danger, nearby threats, lying in ambush, that, at any moment, any second, or any hour of the day, may snatch us away to the world of the dead, the ultimate fate of all destinies. If, in your bones, there is a primal shiver – a forewarning from small signs, of a giant mass of things to come, take heed. Take all necessary precautions, make sure, make double sure. Forget or leave out nothing, overlook no possibility. Leave no margin for error, leave nothing to chance.

For to avoid the great river of death – to survive, endure, and cheat death – we must be ever vigilant, ever cautious, and wary, mindful that the struggle between life and death, in its countless and unfathomable ways, marches on.

My hope is that, with the advances in science and computer modeling, we will find the answers to these problems and learn how to co-exist with the natural forces on Earth and take control of our environmental future, and develop a sustainable model that will carry us far into the future and that's my case for proceeding with caution..

A matter of degrees

The National Center for Atmospheric Research's estimate of how many degrees the average global temperature could increase by 2100 if greenhouse-gas emissions continue to increase:

7.3°
8°
7°
6°
5°
4° 1.4°
3°
2°
1°
0
2010 '30 '50 '70 2100

Source: NCAR

"Frankly, I think we'll regret introducing these organisms into the environment."

MONDAY, FEBRUARY 14, 2005

Climate Icon

A reconstruction of average surface temperatures in the Northern Hemisphere was the highlight of a 2001 U.N. report. The graph's hockey-stick shape is cited as evidence that fossil-fuel emissions are warming the planet.

Difference in temperature compared to 1961-1990 average

Data from tree rings, corals, ice cores and historical records (1000-1980)
Range of uncertainty
Data from thermometers (1902-1998)

+0.75 °C
+0.50
+0.25
0
-0.25
-0.50
-0.75
-1.00
-1.25

1000 A.D. 1100 1200 1300 1400 1500 1600 1700 1800 1900 2000

Sources: Intergovernmental Panel on Climate Change; Michael E. Mann, University of Virginia

1.

Timely conversion to a mix of alternative fuels. Tax incentives to drive Hybrids, Hydrogen or Electric Automobiles. Nuclear energy for power plants instead of coal Fireplants.
Scientists watch atmosphere change closely. Strict Population and emission Controls soft landing. 3 to 5 degrees warmer

3.

Arctic thaw continues unabated. Increasingly warmer Ocean temperature. Massive release of methane into atmosphere. Temperature spike to over and above projected modeled heat levels.
Eco-catastrophe – massive loss of life - 7.5 to 12 degrees higher.

Christian

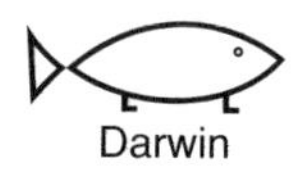

Darwin

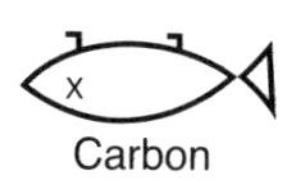

Carbon

2.

Stay the course – political, energy and oil industry resistance to change.
Keep fossil fuels at 70 to 80% usage. Coal Fire plants continue to be the way to fuel expanding energy needs. Population Increase to 9 billion. Hard landing - 5 to 7.5 degrees warmer.
Millions perish.

4.

Ozone depletion – atmospheric mix, deterioration
More CO2 and Methane. Less oxygen – continued increase in global temperature - 12 to 20 degrees.
Ecocide – massive extinction of 95% of All species - no adaptation.

The Domino Effect

Putting Society at Risk

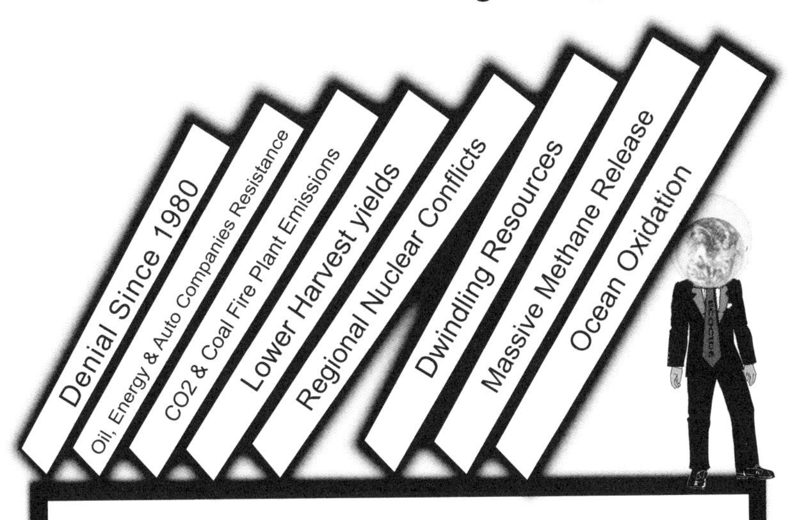

With trillions of dollars of fossil fuels at stake — It is in the best interest of the oil & energy companies to keep the fossil fuel mix high. By ignoring problems that are not a urgent threat today, we put ourselves at greater risk. The good news is: we have control over all but two of the problems; methane release & ocean oxidation. With political, scientific & community will power we could prevent an eco catastrophe.

FOR FURTHER READING

For those who would like to learn about global warming in more detail, I am including here a selection of published articles, statistics, and information from the scientific community and from Web sites that are tracking and analyzing this problem. I gleaned this information from news articles and the Internet, beginning in 1999 when I read that government specialists and a wide range of experts were saying that global warming will reshape the world we live in and will challenge the international community as polar ice caps melt, sea levels rise, and the frequency of major storms grows. This information affected me enough to research and write this book. I hope that what I'm including here will help to make you more aware of the potential problems that our children and grandchildren will have to contend with.

Environmental Outcome

"The Environmental Outcome of Global Warming" -- The Woods Hole Research Center. Go to: www.whrc.org/globalwarming/outcome.htm

Potential Outcome

(The following is a summary of the analysis of potential outcomes of climate change delineated by the Intergovernmental Panel on Climate Change (IPCC) in their second assessment report).

☺ Rising Temperatures

The average surface temperature of the Earth has increased by about one degree F. in the past century. To many, a one-degree temperature change may seem trivial however consider "the year without a summer" -- 1816. That year, atmospheric ash from a volcanic eruption in Southeast Asia decreased solar radiation reaching the Earth's surface and lowered the global mean temperature. As a result, frost occurred in July in New England, and crop failures occurred throughout the world. Yet the temperature change caused by this eruption was less than one degree F. Stommer et al., 1979.

The Earth's surface temperature is projected to increase by 3.5 to 7.4 degrees F. in the 21st century, with the scientists' best guess being about 3.5 degrees F. Scientific modeling suggests that the surface temperature will continue to increase beyond the year 2100 even if concentrations of greenhouse gases are stabilized by that time. If carbon dioxide emissions continue to increase at present rates, however, a quadrupling of the pre-industrial CO_2 concentration will occur not long after the year 2100. Projected temperature increases for such an atmospheric concentration are fifteen to twenty degrees F. above the present-day mean annual global surface temperature, which would take us out of that part of the temperature spectrum that encourages plant and animal procreation.

☺ Sea Level Rise

Increasing global temperatures causes the thermal expansion of seawater and the melting of ice caps, which will raise the sea level. Sea levels rose four to ten inches in the twentieth century and are predicted to rise another six to thirty-seven inches in this century. A doubling of the pre-industrial CO_2 concentration (to 550 ppm) is predicted to result in a sea level rise of greater than forty inches. A sea level rise of eighty inches is projected for an atmospheric CO_2 concentration of 1,100 ppm, a quadrupling of pre-industrial levels. Rising sea levels increase the vulnerability of coastal populations to flooding and cause the loss of land to erosion. Worldwide, some forty-six million people are vulnerable to flooding from storm surges. With a 50 cm sea level rise (approximately 1.5 feet), that number will increase to ninety-two million. Raise sea level one meter (about three feet), and the number of vulnerable people becomes 118 million. A one-meter increase in sea level will be enough to flood 1 percent of Egypt, 6 percent of the Netherlands, 17.5 percent of Bangladesh, and 80 percent of the Majuro Atoll in the Marshall Islands. Rising waters could force the occupants of many small island nations to migrate elsewhere. Since many such places lack the coastal defense systems to cope with higher water levels.

☺ Intensification of the Hydrologic Cycle

Warming will likely increase the amount of fresh water exchanged among the oceans, atmosphere, and land. Rising rates of evaporation will likely result in drier soils. An accelerated hydrologic cycle means greater amounts of precipitation in some areas and will probably result in more frequent and severe storms. Paleoclimatic data suggest that the collapse of the Mesopotamian Empire about 4,200 years ago (2,200 B.C.) corresponds to a sharp cooling event. Alley and DeMenocal, 1998.

☺ The Effect on Fisheries

Great pressure has already been placed on commercial fishing industries, and many of the commercially fished species are in steep decline. Rising temperatures are not predicted to change the global average production; however, significant regional changes are likely. Production is projected to increase in the higher latitudes, in freshwater, and in aquaculture operations. Warmer climates should see an extension of the growing season, a decrease in the natural winter variability, and improved growing rates in higher-latitude regions. These beneficial results, however, may be offset by changes in reproductive patterns, migration routes, and ecosystem relationships.

☺ Food Production

Total global food production is not expected to change substantially as a result of climate change, but production will probably change dramatically regionally. Some areas will see increased crop yields while others will experience a decline, especially in tropical and subtropical regions. The flexibility in crop distribution (the variety of crops that can be grown in a region) is predicted to decline. Developed countries may be able to adapt to these circumstances, but developing countries that struggle with these issues now will suffer even more.

☺ General Comments

1. The warming of the Earth is expected to be rapid, more rapid than any climate change that we know about in the recent history of the Earth, including glacial periods (.5 to one degree C. or more per decade for the middle to high latitudes). The warming is expected to be indefinite in duration. We appear to be entering a period of continued warming. Dr. George M. Woodwell, 1989.

2. The climate is going to be in continual change, [National Center for Atmospheric Research] NCAR's Trenberth said. "I'm not sure people realize this. Inability to plan [for stable weather patterns] may be worse than the changes themselves." –Newsweek, 1997.

3. The Paleoclimatic record suggests that the climate system can respond to various weather conditions, forcing change in a nonlinear manner. For example, approximately 11,500 years ago the Earth warmed by nine to eighteen degrees F. in less than a decade. If humans change the composition of the atmosphere significantly enough, the possibility exists that an abrupt climate shift with substantial social and ecological consequences could occur.

☺ Deserts

The climate of desert regions is likely to become more extreme and even hotter than it is now. The process of desertification will be more likely to become irreversible because of drier soils and land degradation caused by erosion and compaction.

☺ The Cryosphere

In the next hundred years, between one-third and one-half of the world's mountain glaciers could melt, with negative effects on the water supply to rivers, hydroelectric dams, and agriculture. Already being observed in Alaska is the actual extent and depth of the decline of permafrost. The loss of permafrost has adverse effects on native animal species and on man-made infrastructure. A decrease in the extent and thickness of sea ice will likely improve the navigability of the Arctic Ocean, although these losses are already producing negative effects on animal life (the drowning of polar bears, for instance).

☺ Mountain Regions

Warming temperatures will probably induce a shift in the distribution of vegetation to higher elevations, some instances of which have been noted already. Animals that exist only at high elevations may become extinct because of the disappearance of habitat and/or the decline in migration potential. Recreational industries (the snow-skiing industry, for example) are likely to be disrupted, which will impact the economy of some areas. The high-elevation populations of developing nations will probably suffer from a decline in the abundance of food and fuel.

☺ Lakes, Streams, and Wetlands

Climate change is predicted to alter water temperatures, flow patterns, and water levels. Such changes will likely cause an increase in biological productivity in the higher latitudes, but may result in the extinction of some cool- and cold-water species in the lower latitudes. An increased variability in flow, which will increase the frequency and duration of major floods and droughts, will tend to reduce water quality, biological productivity, and the habitat in streams.

☺ Coastal Systems

Climate change and rising sea levels, or changes in storms or storm patterns or storm surges could cause the erosion of shorelines and associated habitat; an increase in the salinity of estuaries and freshwater aquifers; a change in tidal ranges in rivers and bays; a change in sediment and nutrient transport; a change in the pattern of chemical and microbiological contamination in coastal areas; and an increase in coastal flooding. The ecosystems at greatest risk are saltwater marshes, mangrove ecosystems, coastal wetlands, coral reefs, coral atolls, and river deltas.

☺ Oceans

Increased atmospheric temperatures will change patterns of ocean circulation, vertical mixing, wave climate, and quantities of sea-ice cover. These changes will affect nutrient availability, biological productivity, and the structure and function of marine ecosystems. Paleoclimatic (past climate) data and models show that major changes in ocean circulation can be caused by freshwater additions to the oceans from the movement and melting of sea ice or ice sheets and could result in rapid and dramatic changes in climate if the North Atlantic current slows or shuts down completely. Abrupt climatic shifts have caused adverse events that affected human civilizations before, including droughts, devastating floods, and mini-ice ages. This prediction has gained attention already, as in the film The Day After Tomorrow. In the early 1990s, two 100-year floods occurred in less than five years in the Midwestern United States.

Significant changes are occurring in the oceans as the ice melts in Greenland and the arctic, as tremendous inflows of fresh water come from rivers in Siberia, and as mountain snowmelt finds its way into the seas, further freshening and desalinizing the oceans of the world.

☺ Health Effects

A warming Earth will most likely produce a whole spectrum of largely negative impacts on human health. The predicted decrease in the difference between day and night temperatures will result in more thermal extremes. Therefore, an increase in mortality from heat stress is likely (e.g., 465 deaths in Chicago during the summer of 1995). Because of warming, the area of the Earth's surface experiencing "killing" frosts will probably decline. As a result, we will probably experience an increase in the geographic range of vector-borne (e.g., mosquito-carried) diseases such as malaria, dengue, yellow fever, and encephalitis. Currently, forty-five percent of the world's population live within the zone of potential malaria transmission. With predicted temperature increases, we could see an additional fifty to eighty million cases of malaria worldwide, bringing the percentage of the global population living within the susceptible zone to sixty percent.

Rising temperatures will also cause a decline in air quality because of the increased abundance of air pollutants, pollen, and mold spores. Many more cases of respiratory disease, asthma, and severe allergies will no doubt follow. Also, the change in the frequency and intensity of extreme weather events (e.g., storms, floods, and droughts) combined with warmer atmospheric temperatures, will probably create a host of adverse health effects, some of them caused by contaminated water supplies, and death from diseases.

☺ Dramatic Effects on Ecosystems

Both plant and animal species are sensitive to climatic changes. Because of global warming, ideal temperature and precipitation ranges suitable for present life forms may shift dramatically and rapidly, more quickly than some species can adapt to naturally. A decline in biodiversity in most ecosystems is a likely result. A lengthening of the growing season, however, is also predicted for some higher-latitude regions, which means that these areas will probably experience an increase in their agricultural potential.

☺ Forests

Within the next hundred years, many forest species may be forced to migrate between one hundred and 340 miles toward the poles. The upper end of this range is a distance typically colonized by migrating forests over the course of millennia, not decades. A decline in species composition is predicted, and some tree varieties may disappear from the Earth, while new ones may be established.

☺ Rangelands

Changes in growing seasons and shifts in the boundaries between grasslands, forests, and scrublands are among the projected results of changing temperatures and precipitation regimes. Increased levels of carbon dioxide in the atmosphere may also cause a decline in the food values of grasses for herbivores.

☺ Scientific Evidence

Through the study of ancient ice core samples from Antarctica, scientists have determined both the concentration of carbon dioxide in the atmosphere and the Global Mean Annual Temperature for the past 160,000 years. By examining the graph of Global Mean Annual Temperature and Atmospheric Carbon Dioxide Concentration over this time period, researchers have established a direct relationship between the two levels. In fact, the fluctuations in one plot seem to be, for the most part, mirrored in the other. Also, the amount of carbon dioxide in the atmosphere and the global temperature have been increasing since the end of the last ice age, about ten thousand years ago.

Why then are the more recent increases of such concern? First, because the latest increases are occurring at rates that have not been observed since the last ice age (Intergovernmental Panel on Climate Change, 1995) and have been observed before only in association with dramatic shifts in climate. Second, the dramatic increase in the carbon dioxide concentration in the atmosphere during the past 150 years (from about 280 parts per million to about 360 parts per million) is largely an anthropogenic (human-caused) effect (IPCC 1995)

After several years of investigation and consultations with thousands of scientists, the Intergovernmental Panel on Climate Change--a UN task force that is examining the plausibility of human-induced climate change--concluded that "the balance of evidence suggests a discernible human influence on global climate." In addition to the extensive investigation and clear conclusion by the IPCC, a letter signed by 2,600 scientists was submitted to President Clinton in the summer of 1997. The letter urged the United States to take a leadership role in reducing greenhouse gas emissions and preventing the onset of even more intense and continuous global warming.

As predicted by the reports of the IPCC, the climate has indeed been changing. The ten hottest years in the past century have all occurred since 1980, with 1990, 1995, and 1997 being the hottest years on record until 2004 and 2005 tied for that distinction. Some 465 people died from heat-related causes in Chicago during the summer of 1995, and three hundred people died from the heat in India that same year (Gelbspan, 1997).

Earlier that year, a giant block of ice the size of the state of Rhode Island broke off the Larsen ice sheet in Antarctica. The melting of the Antarctic ice sheet is an event that has been long predicted by climatologists as an indication of a warming atmosphere (Gelbspan, 1997).

The temperature data for 1998 are staggering. Data collected by NOAA (National Oceanic and Atmospheric Administration) indicate that 1998 was by far the warmest year in recorded history. The global mean surface temperature in 1998 was .66 degrees C. (1.20 degrees F.) above the long-term (1880-1997) average value of 13.8 degrees C. (56.9 degrees F.). A particular year may be the warmest year on record but not the warmest on record for any single month. For nine of the twelve months of 1998, however, the global average temperature exceeded the monthly records for all previously recorded years. In other words, in the one hundred-plus years that temperature data have been recorded, we have never recorded a warmer January, February, March, April, or May, June, July, August, or October than in 1998. In addition, it's worth noting that the previous monthly records had all been established within the preceding ten years, between 1988 and 1997.

Throughout 1998, we heard reports about the severe weather that occurred around the world as a result of El Niño. This is an atmospheric phenomenon that has been observed for hundreds of years, but what is often unstated is that the frequency and intensity of El Niños are apparently increasing. Scientists once believed that El Niño conditions occurred once every five to seven years. Now they seem to be occurring every three to five years (Showstack, 1998). "The 1990 to mid-1995 persistent warm phase of the El Niño Southern Oscillation was unusual in the context of the last 120 years. (IPCC, 1995)

Does a clear link exist between El Niño activity and a host of other recent, relatively short-term climatic events and the amount of greenhouse gases that humans have spewed into the atmosphere in the past 150 years? No. But have climatic events unprecedented in the human experience occurred in recent times? Yes. Will the direct cause-effect link be evident before it's too late? Probably not. So, in the face of uncertainty, do we hesitate to act? Or, considering the potential outcomes, do we act? (See Dr. John Holdren's address at the White House Conference on Climate Change on October 6, 1997.)

☺ The Culprits

While the concentrations of almost all greenhouse gases have been increasing since the beginning of the Industrial Revolution, carbon dioxide has had the greatest effect on changing the climate. During the 1980s, humans released 5.5 billion tons of carbon dioxide into the atmosphere annually by burning fossil fuels (coal, oil, natural gas) for heat, transportation, and electricity. An additional 1.6 billion tons were released from anthropogenic (human-induced) changes in land use (i.e., clearing land for agriculture, pastures, etc.), mostly through deforestation in the tropics.

Where do those 7.1 billion tons of atmospheric carbon go? Ocean modelers find that the sea absorbs about two billion tons a year. Another two billion tons are taken up by a presently unidentified "sink" or reservoir of carbon (see The Missing Carbon Sink). This leaves 3.1 billion tons of CO_2, an amount that global atmospheric measurements indicate is simply being added to existing concentrations in the atmosphere. The result is that this concentration of CO_2 is increasing at a rate of approximately 1.5 ppm (parts per million) per year. Since the beginning of the Industrial Revolution it has increased by about thirty percent.

These flows or "fluxes" within the Global Carbon Cycle may be summarized by using the formula Atmosphere = Fossil Fuel emissions + Land Use Change — Ocean Uptake — Missing Sink.

Human beings are causing the release of carbon dioxide and other greenhouse gases into the atmosphere at a rate that is much greater than the Earth can absorb and recycle. Fossil fuels--oil, coal, natural gas, and their derivatives--are formed by the compression of organic (once living) material for millions of years. We are burning billions of tons of these fuels each year. Why is this disconcerting? Because the CO_2 expelled into the atmosphere by our activities does not disappear immediately or even in a short time. Greenhouse gases remain in the atmosphere for many decades or even centuries. This means that the CO_2 we emit today will most likely continue to affect the climate well into our children's future and probably into the future of our grandchildren.

Despite the widespread recognition of this fact, global emissions from fossil fuels consumption continue to increase at a rate of about one percent a year (IPCC, 1995). These emissions will increase even more as the developing world becomes more industrialized. As of 1995 the industrialized world (the United States, Western Europe, Eastern Europe, and the former Soviet Union) contributed more than seventy percent of the total emissions. If the use of fossil fuels continues to increase at present rates, by 2035 humans will be releasing twelve billion tons of CO_2 into the atmosphere annually--about half from the developed nations and half from the developing countries (IPCC, 1995).

What the Skeptics Don't Tell You

Paul Ehrlich has said, "Laypeople frequently assume that in a political dispute the truth must lie somewhere in the middle, and they are often right. In a scientific dispute, though, such an assumption is usually wrong."

It is human nature to protect our own interests. We may recall the extensive and incredibly successful campaign of the

American tobacco companies to conceal the link between cancer and the use of tobacco products. For decades they knew the reality of the addictive nature of nicotine and the carcinogenic effects of tobacco use. But they successfully kept that reality hidden from the American public for all that time.

The oil, coal, gas, and mining industries stand to lose tremendously if the truth about global warming becomes accepted by American society. Just as the tobacco industry invested millions in keeping its deadly secret, these industries have attempted to hide and discredit the link between CO_2 emissions and a warming Earth. They have funded, promoted, and used as witnesses a handful of greenhouse skeptics who have widely and loudly proclaimed that global warming is a myth.

The truth is, we now have a scientific consensus, for the most part, that says human action is causing a warming of the earth. Scientists who subscribe to this view recognize, nonetheless, that the specific outcome of the warming is uncertain. The oil and mining industries have chosen to focus on the word uncertainty in their sound bites and media campaigns. Regrettably, they have disingenuously sought to extend the notion of uncertainty from the outcome of global warming to the global warming phenomenon itself--a flagrant and unjustifiable attempt to discredit what we already understand.

The greenhouse skeptics, for the most part, fail to submit their work to the process of peer review. This is the process by which others in the scientific community doing similar work are asked to comment on, criticize, and replicate a scientist's findings before they are published. Peer review is an integral and established part of sound science. If a paper is not peer-reviewed, there is no verification of the credibility or validity of the science being undertaken.

Several members of the U.S. Congress have overlooked the importance of this process, and many have given equal, if not more, credit to the statements of non-peer-reviewed reports. The degree to which the greenhouse skeptics and their organizations have succeeded in blurring and undermining the facts about global warming is evident when we examine the statements of several congressmen about the issue. For example, in response to a statement made by David Gardener from the EPA regarding the potentially catastrophic and irreversible effects of sea-level rise, Dana Rohrbacher, a California Republican, surfer, and member of the House Science Subcommittee on Energy and the Environment, flippantly remarked, "I am tempted to ask what this will do to the shape of the waves and the ride-ability of the surf. But I will not do that. I'll wait until later when we get off the record." (Gelbspan 1997).

Widespread efforts are being made to discount the scientific evidence, as illustrated by the following example. In early 1998, a senior scientist at The Woods Hole Research Center received a letter sent by an organization identified only as "GWPP." Enclosed were several documents: a short note signed by Frederick Seitz, former president of the National Academy of Sciences; a photocopied article that had appeared in the Wall Street Journal on December 4, 1997, entitled "Science has Spoken: Global Warming Is a Myth," a scientific paper entitled "Environmental Effects of Increased Atmospheric Carbon Dioxide," and a petition to be signed and returned. The note, which was conspicuously lacking any sort of header or organizational letterhead, urged the reader to sign the enclosed petition and proclaimed that, "Research data on climate change do not show that human use of hydrocarbons is harmful. To the contrary, there is good evidence that increased atmospheric carbon dioxide is environmentally helpful." The petition went on to state:

We urge the United States government to reject the global warming agreement that was written in Kyoto, Japan, in December of 1997, and any other similar proposals. The proposed limits on greenhouse gases would harm the environment, hinder the advance of science and technology, and damage the health and welfare of mankind.

There is no convincing scientific evidence that human release of carbon dioxide, methane, or other greenhouse gases is causing or will, in the foreseeable future, cause catastrophic heating of the earth's atmosphere and disruption of the earth's climate. Moreover, there is substantial scientific evidence that increases in atmospheric carbon dioxide produce many beneficial effects upon the natural plant and animal environments of the Earth.

The "scientific" paper looks like any other reprint from a scientific journal. What the authors fail to mention, however, is that this article was neither peer-reviewed nor previously published in any shape or form. To anyone not intimately familiar with scientific papers, this article would probably help to perpetuate the myth that there is still uncertainty about the warming of the Earth. This paper was in part produced by employees of the George C. Marshall Institute. This organization, founded in 1980 to issue reports promoting President Reagan's "star wars" defense program, conducts no original research. Most recently, it has focused on issuing reports dismissing climate change (Gelbspan, 1997). Disseminating misinformation, spinning the facts of science, and creating controversy in order to meet the goals of political parties and energy companies have been common tactics since the 1970s, and it's much worse today.

Many civilizations have imploded at the height of their power. Civilizations that ignore their environmental limits have disintegrated, as chronicled by Pulitzer Prize-winner Jared Diamond in his best-selling book Collapse. Mr. Diamond has studied the cultures of the world and found that some societies make disastrous decisions, the most common being the failure to perceive a problem when it is a slow trend concealed by widespread, up-and-down fluctuations. Global warming is a classic example of a creeping problem with extensive implications and fluctuating weather patterns, but instead of affecting one civilization it will affect all civilizations and all living things on the planet. Slow trends conceal dangerous environmental variations. It is very difficult to recognize a slow, devious problem that each year is slightly worse than the year before, although it may be detected over a span of twenty years. I have observed some changes in the weather myself in the last thirty years, as many of you have, I'm sure, such as milder or harsher winters, depending on where you live, the west or the east; more exotic storm cycles; more droughts; and the earlier than usual arrival of spring. The human sense of normalcy changes gradually and almost imperceptibly, so it may take a few decades of slight year-to-year change before people realize with a great jolt that a major problem is upon us. Gradual change can be a terrible threat if it's leading in the wrong direction from our goal of survival.

Rational behavior, as defined by behavioral scientists, is behavior that some people or groups of people use to advance their interests at the expense of others. Behavior like that is "rational" because it employs correct reasoning and may even be backed by law and encouraged by society. The worst perpetrators reap huge profits that are concentrated in the hands of a few individuals or corporations. Loss, on the other hand, is distributed among many people, and since each person is losing only a small amount, the many have little motivation to join forces and fight a major battle over it. Therefore, the minority interest successfully undermines and sabotages the interest of the majority.

The "tragedy of the commons," a theory proposed by an English professor, explains this by observing the logic of collective actions. For instance, consider a situation where many consumers are using or harvesting a communally owned commodity such as fish in the local area of the ocean like Gloucester, Massachusetts, which has had a viable fishing industry for over four hundred years. Or consider a communal pasture for sheep. If everybody overharvests a resource, it will become depleted, then slide into a steep decline or even disappear completely. This happened with the cod fishing industry in Gloucester, which has virtually disappeared, when at one time million-pound hauls were common. At the time of this writing, all commercial fishing has been restricted to the point where the industry has been thrown into a death spiral. This has affected not only the fishermen, but every business from hardware to clothing. A four-hundred-year-old way of life is disappearing. What will replace commercial fishing when this industry collapses altogether?

The reasoning that leads to such a collapse goes like this: "If I don't catch that fish or let my sheep graze in that pasture, some other fishermen or sheep herder will, so it makes no sense for me to refrain from overfishing or overharvesting." The "rational" behavior is to get it or harvest it before the next person can, even though the destruction of the community resource brings harm to all the citizens, without strict government regulations. We can point to innumerable examples of overuse that have led to the loss of forests, fisheries, and wildlife and to the extermination of large mammals, birds, and reptiles, from the buffalo to the codfish, everywhere humans have settled in the last fifty thousand years.

The reasons for the collapse of a civilization are many and varied, but the primary causes are the overuse of their natural resources and climate change. The original residents of Easter Island cut down all their forests, which contributed to soil erosion and a decline in their agriculture, eventually leading to famine, war, cannibalism, and the destruction of their society. The Anasazi Indians of the Southwest, the Mayans, and the Greenlanders, likewise suffered a similar fate when their environment changed because of a climatic shift and droughts, which led to deforestation and the loss of other key natural resources.

Today, wherever we have problems with overpopulation and environmental stress--places like Afghanistan, Bangladesh, Haiti, Madagascar, Nepal, Rwanda, the Philippines, Iraq, Saudi Arabia, and Somalia--we also see hot spots for terrorism, political upheaval, and genocide. In countries like these, where most of the population are desperate, undernourished, and without hope, the people blame their government and eventually their religious leaders, and then they try to flee to another country at any cost. Look at the situation along the border between Mexico and the U.S., for example. People who are desperate to find a new life may become terrorists, start civil wars, or support terrorism, revolution, or genocide. Overwhelming poverty is the key precursor to war and government failure. Global warming and diminishing oil and water supplies could lead to a world war, this one with a number of new countries wielding the atomic bomb.

The outcome of a collapse of the world's civilizations that is caused by nuclear war or a rapid climate change is too nightmarish for most of us to comprehend. We just can't imagine it. That is why the slow, gradual problem of global warming is so insidious. Then add to the equation the problems of diminishing resources of oil and fresh water and the challenge of supporting eighty-five million more people each year. We should also factor in the influence of all the Christian fundamentalists who believe in the imminent arrival of the prophesied End of Times and therefore see no reason to conserve anything. Why save anything for future generations, they say, since no one will be here anyway? In addition, we have the mounting potential for conflict between the Christian world and the Muslim one, the latter of which will outnumber the first by the millions by 2025. Many of them, we must keep in mind, live in countries controlled by a megarich elite, where much of the populace struggles to survive and where a growing number of them will do anything to further their fanatical causes.

Take all these ingredients and try to imagine the possibilities for the future. As I have said before, the picture is not a pretty one.

APPENDICES

APPENDIX 1

In the News

Press Release from Milan, Italy (via Internet)

November 27, 2003;

Billions likely to suffer Water shortages from melting glaciers

Milan, Italy -- Unless governments take urgent action to prevent global warming, billions of people worldwide may face severe water shortages as a result of the alarming melting rate of glaciers, warned today a head of the 9th Conference of the Parties (COP9) to the United Nations Framework Convention for Climate Change (UNFCCC).

A new World Wildlife Federation (WWF) paper on climate change and global glacier decline shows that increasing global average temperatures in the coming century will cause continued widespread melting of glaciers, which contain 70 percent of the world's freshwater reserves. An overall rise of temperature of four degrees Celsius before the end of the century would eliminate almost all of them. British scientists are predicting a warming of 5.4 degrees on average by the end of the century.

The melting of glaciers will lead to water shortages for billions of people, as well as sea level rises threatening and destroying coastal communities and habitats worldwide. According to the conservation organization, the regions most at risk from melting glaciers due to climate change are:

- ☺ Greenland and the Antarctic.
- ☺ Ecuador, Peru, and Bolivia, where glaciers supply water all year round, and are often the sole source of water for major cities during dry seasons.
- ☺ The Himalayas, where the danger of catastrophic flooding is severe and glacier-fed rivers supply water to one-third of the world's population
- ☺ Small island nations such as Tuvalu, where sea level rise is submerging lowlands and saltwater is invading vital drinking water supplies.

"These glaciers are extremely important because they respond rapidly to climate change, and their loss directly affects human populations and ecosystems," said Jennifer Morgan, director of WWF's Climate Change Programme. "WWF expects developed country ministers at the upcoming COP9 meeting to clearly demonstrate what their countries are doing now to combat climate change and commit to deep reductions in the future to prevent such catastrophes."

WWF is urging the governments of more than 180 countries attending the UNFCCC COP9 from 1-12 December 2003, in Milan to speed up in taking action to combat climate change. WWF also calls on all countries to ensure that Russia ratifies the Kyoto Protocol as soon as possible. In the negotiations, WWF is placing top priority on the conclusion of strong rules for forests and land-use activities that absorb carbon from the atmosphere in the Clean Development Mechanism at COP9. In addition, WWF is encouraging ministers from industrialized countries to describe how their governments are implementing measures to meet Kyoto targets and commit to deeper emission cuts in the near future.

News Article dated January 17, 2004
(via Internet)
Salmon "cooking" in Scottish rivers

Aberfeldy, Scotland -- Last year, Scotland's hottest year on record, saw hundreds of adult salmon die in rivers across Scotland. The World Wildlife Federation (WWF)-Scotland warns today that, if climate change patterns continue, 2004 will see a rise in salmon death unless urgent action is taken. The warning was made at the Kenmore Salmon Seminar in Perthshire, Scotland.

The Atlantic salmon (Salmo salar) is becoming a species under threat from global warming. A coldwater species, the fish return from the sea to spawn in Scottish rivers, traditionally cold, clean waters. Rising river temperatures are potentially fatal to adult salmon because, as cold-blooded creatures, their metabolism is intrinsically linked to the water temperature. As the river warms up the salmon need more oxygen, but as the water heats up, their ability to extract enough oxygen from the water decreases. Independent research has shown that 20 degrees Celsius is the lethal threshold for the wild salmon in Scottish rivers.

"River temperatures are set to rise if predictions about climate change come true this year," warns Mike Donaghy, Freshwater Policy Officer for WWF-Scotland. "If this level of temperature is sustained in Scotland, it will be lethal for our already vulnerable wild salmon populations."

With Scotland reaching air temperatures of more than 20 degrees Celsius many times in 2003, and with six of the ten hottest years all being in the past decade, the climate change pattern looks set for temperatures to rise--and alongside it, salmon deaths. WWF-Scotland believes that intervention is needed by anglers, riverbank owners, farmers, and fisheries alike to ensure a healthy future for the fish.

WWF-Scotland proposes:

- Reducing the amount of water being taken from rivers for domestic, industrial and recreational use and allowing water levels to recover.
- Planting exposed river banks with tree cover from native trees, thereby shading the river and cooling the water.
- Eliminating pollution from farm slurries and fertilizers.
- Agreeing to a voluntary code to halt all angling when water temperatures are above 20 degrees Celsius.

Global Warming vs. Our Economic Health

By John Buell

(published July 31, 2001 by Bangor Daily News)

Last week, I was struck by a curious coincidence. On the day the Bush administration condemned the rest of the world for global warming agreements, the Environmental Protection Agency issued another insistent warning on smog for coastal Maine.

While the president worries that curbs on greenhouse gases will harm the economy, our growing dependence on fossil fuels is already a major contributor to costly public health problems. Even if the greenhouse effect were a figment of rabid environmentalists' imaginations, there are immediate reasons to be concerned about the ways we use energy.

The message from the EPA was blunt and unsettling. Air quality was "predicted to be unhealthy along coastal Maine due to elevated concentrations of ground-level ozone, commonly called smog. Anyone can be affected by air quality, but groups particularly sensitive include children and older adults who are active outdoors, and people with respiratory disease such as asthma. Sensitive people who experience effects at lower concentrations are likely to experience more serious effects at higher concentrations. Still, even the healthiest people may find it difficult to breathe when smog levels are very high."

It went on to suggest that: "All people, especially children, should limit strenuous outdoor activity during the afternoon and early evening hours, when ozone levels are highest."

Those who anticipate long-term climate change point out that more than half of the threat derives from carbon dioxide released by burning fossil fuels. But regardless of the climatic effects of fossil fuel combustion, immediate damage is being done. Ground-level ozone also forms during the warm weather when pollution from vehicles, industry and power plants bakes in the hot sun.

The American Lung Association points out that 100 million Americans live in communities where ozone concentrations regularly exceed federal standards. Though the causes of asthma are poorly understood, there is little doubt that high concentrations of smog exacerbate the condition. With asthma ranking as one of our fastest-growing chronic health problems, smog itself is heavily implicated in our escalating health costs.

Many tourists come to coastal Maine both for recreational activities and to escape the [poor] air quality experienced in many Northeastern cities. How surprised they--and we--are to learn that even Acadia National Park can hardly escape the consequences of deteriorating air quality. Prevailing upper-air patterns make coastal Maine the end of the tailpipe for a toxic brew emanating from the Northeastern metropolises.

APPENDIX 2

Time Scales of Processes Influencing the Climate System

As determined by the Intergovernmental Panel on Climate Change (IPCC) in the Second Assessment Report, 1995

Stabilization of atmospheric concentration of long-lived greenhouse gases given a stable level of greenhouse gas emission	Decades to millennia
Equilibrium of the climate system given a stable level of greenhouse gas concentration	Decades to centuries
Equilibrium of sea	Centuries
Restoration or/rehabilitation of damaged or disturbed ecological systems = level	Decades to centuries

All of that is accurate given a stable climate.

Some changes are irreversible (e.g., species extinction). Some disturbed ecosystems are likely to disappear forever. We must no longer think of human progress as a matter of imposing ourselves on the natural environment. The world--the climate and all living things--is a closed system; what we do has consequences that eventually will come back to affect us.--

-- UNEP

APPENDIX 3

The Kyoto Protocol

Background

In 1987, the Intergovernmental Panel on Climate Change (IPCC) was formed by the United Nations Environmental Programme (UNEP) and the World Meteorological Organization (WMO), "to assess the available scientific, technical and socioeconomic information in the field of climate change." In 1990, the first report of the IPCC was released. It called for immediate action to avoid the effects of a warming climate. This report was supported by representatives at the Second World Climate Conference, which occurred later that same year. Immediate negotiation of a framework convention on climate change was called for by the representatives of this second climate conference. The UN General Assembly created a committee to draft a treaty for the upcoming Earth Summit in Rio de Janeiro. That treaty, now known as the United Nations Framework Convention on Climate Change (UNFCCC), was subsequently accepted and signed by more than 150 nations represented at the Rio Conference.

The ultimate objective of the UNFCCC is: 'Stabilizing greenhouse gas concentrations in the atmosphere at a level that would prevent dangerous anthropogenic interference with the climate system." It goes on to state that "such a level should be achieved within a time-frame sufficient to allow ecosystems to adapt naturally to climate change and to insure that food production is not threatened and to enable economic development to proceed in a sustainable manner."

Countries ratifying the convention agreed:

- To develop programs to slow climate change.

- To share technology and cooperate to reduce greenhouse gas emissions.

- To develop a greenhouse gas inventory listing national sources and sinks.

At the Earth Summit, it was generally agreed that the responsibility falls upon the developed nations to lead the fight against climate change, as they are largely responsible for the current concentrations of greenhouse gases in the atmosphere. The original target for emission reductions that was generally accepted in 1992 was that the developed nations should, at a minimum, seek to return to 1990 levels of emissions by the year 2000. Additionally, developed nations should provide financial and technological aid and assistance to the developing nations to produce inventories and work toward more efficient energy use. The parties to the convention agreed to convene again in Kyoto, Japan, in 1997 to implement legally binding agreements on greenhouse gas emissions.

There are inherent conflicts of interest related to the issue of climate change. Traditional points of digression between developed and developing nations of the world become overwhelmingly apparent during climate change negotiations. The developed world has a relatively high standard of living in comparison to the developing world. The developed world is largely responsible for the current dangerous levels of greenhouse gases in the atmosphere, yet the developing world will likely be hit the hardest by the outcomes of climate change.

Concern about the rates of population growth and future industrial growth in developing nations has caused industrialized nations to demand that developing nations be bound by any agreement on emissions reductions. The developing nations argue that they don't possess the economic or technological resources to buy into an agreement yet. They see the demands of the developed nations as an attempt to stifle their economic and industrial growth, while they are desperately striving for a higher standard of living and a better life. They ask why they should be responsible for remediating a mess they did not create.

In the United States, reservations about the lack of commitment by developing nations led to the passage of the Byrd-Hagel Resolution, passed by the U.S. Congress in early 1997. The resolution had two main points:

1. The U.S. will not enter into an agreement to reduce greenhouse gas emissions that will be detrimental to the economy of the U.S.

2. The U.S. will not enter into an agreement to reduce greenhouse gas emissions that does not require "meaningful involvement" on the part of developing nations.

While this resolution was an effort to safeguard U.S. economic interests, it has been a significant psychological and legal impediment to stringent restrictions on greenhouse gas emissions. Such obvious reservations about emissions reductions on the part of the world's richest and most powerful nation does not foster optimism about the likelihood of an aggressive international agreement to curb climate change.

The Kyoto Conference

In December of 1997 the countries that had met in Rio convened in Kyoto, Japan, to develop a set of legally binding agreements on the reduction of greenhouse gas emissions. Prior to the conference, several developed nations had made proposals outlining the extent to which reductions should take place. The U.S. proposed that nations should be required to stabilize their greenhouse gas emissions at 1990 levels in the interval of 2008 to 2012. (Keep in mind that this is eight to twelve years later than was proposed as a minimum target in Rio.) The European

Union proposed that nations should be required to reduce their emissions to 15 percent below 1990 levels by the year 2010.

Kyoto was not just a meeting of delegates sent by each nation to discuss and draft a greenhouse gas reductions agreement, but rather it was a collection of representatives from every organization with a vested interest in the outcome of the agreement, from lobbyists for oil and coal corporations to the directors and chairmen of NGOs (nongovernmental organizations) like Greenpeace and the World Wildlife Fund to ecologists and climatologists studying the issue of warming to the handful of greenhouse skeptics to numerous representatives from the U.S. Congress. The stakes in this type of agreement are high, and the chasm between the developed and developing nations becomes that much wider and more apparent.

After ten days of discussion and sometimes heated debate, the delegates at the Kyoto Conference reached an agreement. The Kyoto Protocol calls for the reduction of greenhouse gas emissions for several industrialized nations below 1990 levels by 2008-2012. The U.S. agreed to a 7 percent reduction, and the European Union and Japan agreed to 8 percent and 6 percent reductions, respectively. Twenty-one other industrialized nations agreed to meet similar binding targets. The Protocol allows for the trading of "emissions quotas" among industrialized nations, a significant victory for the United States. Emissions trading would allow nations that failed to meet their binding targets to purchase emissions credits from nations that had emissions levels that were lower than their required targets. This would allow a nation like the U.S., which has high emissions levels but also a lot of capital, to satisfy the agreement. However, despite adamant opposition by the U.S. and other industrialized nations (the Kyoto Treaty was subsequently ratified by other nations, including Canada and Russia), the Protocol also indicated that no binding commitments would be required of developing countries like India and those in South America, because this would burden developed countries like the U.S. and those in Europe with the bulk of the cost of implementing the accord.

Ratification: Possible or Impossible?

Although many countries signed the Kyoto accord, the struggle to make this agreement international law has only just begun. Before the Kyoto Protocol can be implemented, it must be ratified by national governments the world over. At least fifty-five parties--nations who ratified the original United Nations Framework Convention on Climate Change (176 nations)--to the convention must ratify the Protocol, and ratifying countries must account for greater than 55 percent of greenhouse gas emissions in 1990 in order for the Protocol to become international law. As of February 2, 1999, seventy-three nations have signed the Protocol (the United States among them), but only three have ratified it--Fiji, Tuvalu, and Trinidad and Tobago. Since the United States and Russia are responsible for 36 percent and 17 percent of 1990 greenhouse gas emissions, respectively, if both nations fail to ratify the Protocol they can bar it from becoming international law. At present, in light of the Byrd-Hagel Resolution and intense cynicism and opposition by many congressmen and industrial leaders, chances of ratification in the U.S. seem bleak.

A delegation from the Clinton administration helped to negotiate the Kyoto Protocol, and now the Bush administration is faced with the challenge of satisfying the requirements laid out in the Byrd-Hagel resolution and then garnering enough support for Senate ratification. But we must keep in mind the Republicans' close ties to the energy sector and Vice President Dick Cheney's closed-door meetings with top oil companies, whose earnings have grown to almost incomprehensible amounts, such as ExxonMobil's $235 billion (and a $36 billion in profit in 2005 alone). Considering these facts, we may have little hope that this administration will rein in fossil-fuel consumption; in fact, President Bush has said several times that he is solidly behind the oil industry and the use of fossil fuels.

Many of the other nations of the world are waiting for the U.S. to take the lead on this issue. What purpose does an international agreement on reducing greenhouse gas emissions serve if the nation that is considered the world's superpower (and is currently responsible for more than 25 percent of the greenhouse gas emissions) doesn't sign on? Can we blame others for their hesitancy? So, while this issue has disappeared from the front page of the newspapers, the greatest battle, at least in the United States, has yet to be waged.

If the Protocol was ratified and implemented by all parties to the Kyoto conference, it would result in a 5.2 percent reduction of greenhouse gas emissions below 1990 levels. Through the implementation of such an agreement, anthropogenic emissions would be reduced from around 7.2 billion tons per year to about 6.8 billion tons per year. From an environmental standpoint, this agreement falls woefully short of measures needed to head off the warming of the earth. Most scientists studying this issue are calling for a stabilization of the composition of the atmosphere. That would mean emissions reductions on the order of 50 percent of present levels in addition to preventing further deforestation. The Kyoto Protocol is seen by most environmentalists as a tiny step in the right direction. If this measure fails to receive the necessary support for ratification, the future for climate change legislation looks gloomy.

Human action is essential, however. As stated by George M. Woodwell, director of The Woods Hole Research Center, "There is at the moment no obvious mechanism that will slow, stop, or otherwise deflect the warming short of stabilization of the composition of the atmosphere by human action."

Three known human actions can help us move toward the stabilization of the composition of the atmosphere:

- Decreasing the use of fossil fuels by switching to renewable energy resources;
- Decreasing or eliminating deforestation; and
- Increasing rates of reforestation

Immediate action is imperative when the time scales of remediation are considered.

APPENDIX 4

Government and Nongovernmental Organizations

Listed below are governmental and nongovernmental organizations that are concerned with various aspects of environmental issues.

Governmental Agencies

- Bureau of Land Management
- Bureau of Reclamation
- Bureau of the Census
- Council on Environmental Quality
- Department of Agriculture
- Department of Energy
- Environmental Protection Agency
- Fish and Wildlife Service
- Food and Drug Administration

Nongovernmental Organizations (NGOs)

- Advocates for Youth
- Air and Waste Management Association
- American Forests
- American Lung Association
- American Rivers, Inc.
- Bread for the World
- Center for Clean Air Policy
- Center for Health, Environment
- Center for Science in the Public Interest
- Chesapeake Bay Foundation
- Clean Water Action
- Common Cause
- Community Transportation Association of America
- CONCERN, Inc.
- Conservation International

- Consumer Federation of America
- Critical Mass Energy Project
- Defenders of Wildlife
- Ducks Unlimited, Inc.
- Earth First! Journal
- Earth Island Institute
- Earthwatch-Expeditions
- Environmental Action Foundation
- Environmental Defense Fund, Inc.
- Environmental Law Institute
- Freedom from Hunger
- Friends of the Earth
- Greenpeace USA
- Habitat for Humanity International
- Institute for Local Self-Reliance
- International Planned Parenthood Federation
- Izaak Walton League of America
- Land Trust Alliance
- League of Conservation Voters
- League of Women Voters Education Fund
- National Audubon Society
- National Park Foundation
- National Parks and Conservation Association
- National Wildlife Federation
- Natural Resources Defense Council
- Oxfam America
- Physic. for Social Responsibility
- Plant Drum Foundation
- Planned Parenthood Federation of America, Inc.
- Population Action International
- Population Connection
- Population Reference Bureau
- Population Resource Center
- Public Citizen
- Public Interest Research Group
- Rachel Carson Council, Inc.
- Rainforest Action Network

- Rainforest Alliance
- Renew America
- Resources for the Future
- Rocky Mountain Institute
- Rodale Institute
- Sea Shepherd Conservation Society
- Social Investment Forum
- Student Conservation Association, Inc.
- The Green Center
- The Ocean Conservancy
- The Population Institute
- The Wilderness Society
- Trust for Public Land
- U.S. Public Interest Research Group
- Union of Concerned Scientists
- Water Environment Federation
- Wildlife Habitat Council
- Wildlife Management Institute
- Wildlife Society
- World Resources Institute
- World Wildlife Fund
- Worldwatch Institute

APPENDIX 5

Some Ideas on Helping to Save the Planet

These are some of the things you can do:

- Decrease the use of fossil fuels by switching to renewable energy sources. Iceland has switched to all hydrogen-powered cars. Make your next car a hybrid or one that runs on ethanol or, even better, one that is hydrogen-powered.

- Decrease or eliminate deforestation. For every tree cut down, one must be planted. Plant a tree or two.

- Increase rates of reforestation. Plants absorb CO_2 from the atmosphere.

- Increase the monitoring and support of natural habitats and all their resident species.

- Support or elect only public officials who have a keen awareness of global warming problems and who back alternate energy sources.

- Talk to your friends and family to see what they know and think about global warming.

- Start doing more environmentally friendly things such as walking and biking.

- Support population control (limit of one or two children per family).

- Stop wasting water in your home. Using artificial turf or natural landscaping will save thousands of gallons of water per year.

- Save energy in your home (contact your power company for tips).

- Join organizations like the Natural Resources Defense Council.

- Find other ways to get involved and encourage family members and friends to do the same thing.

In your home:

- Heating and cooling: The average American household contributes about four tons of greenhouse gas pollution to the atmosphere each year. Individual households produce 21 percent of America's global warming pollution. Heating and cooling equipment uses most of the energy in a home. To lessen the impact you can:

1. Plant trees to shade your house and lower cooling costs in summer.

2. Replace your roof with light-colored or reflective material.

3. Insulate your attic, ceilings, floors, walls, ductwork, and pipes. Doing so can reduce home heating bills by 25 percent.

4. Use a programmable thermostat that heats or cools rooms only when necessary.

5. In winter, set your thermostat to a lower temperature at night and when the house is

unoccupied. You can reduce global warming pollution by 350 pounds annually by lowering the thermostat only two degrees in winter.

6. Help to warm your rooms in winter by opening the shades to let in sunlight.

7. Close your shades in the summer to keep the air cool inside.

8. Buy "green power." Electricity customers in forty-two states can now buy green power through their utility company or an alternative power supplier.

9. Use solar energy. Two hundred thousand American households now use solar energy.

■ Lighting: Electric lights consume about 12 percent of the home-energy dollar and produce more than a ton of carbon dioxide pollution in the average home each year. To decrease those numbers you can:

1. Use sunlight to help light rooms as much as possible by opening your shades.

2. Replace old-fashioned incandescent light bulbs with energy-efficient compact fluorescent lights (CFL). A 38-watt CFL replaces a standard 150-watt bulb. Better than they were in the past, CFLs also last up to fifteen times longer than conventional light bulbs. For every old incandescent bulb that you replace you will reduce global warming pollution by about 1,300 pounds.

3. Turn off lights when you don't need them.

4. Use motion-sensors to turn lights on and off automatically when someone enters and leaves a room.

- Appliances: Refrigerators and freezers are second only to heating/cooling systems as energy-eaters, and other appliances aren't much better. Cut your cost and pollution by:

1. Using only Energy Star products, which offer the best energy savings.

2. Don't buy a major appliance that is either too large or too small for your needs. Both extremes waste electricity--and your money.

3. Use power strips. TVs, VCRs, cable TV boxes, and video game boxes also use energy when they're off--almost as much as when they're turned on. Plug them into a power strip and shut off power to the whole strip when you're not using them.

4. Unplug small appliances. Toasters, cell phone and other kinds of chargers, and even plug-in air fresheners can rack up kilowatt hours when they're not in use.

On the Road

You can lower the cost of operating a vehicle and reduce global warming by following a number of simple guidelines, including:

- Use ethanol for fuel.
- Don't exceed the speed limit. Every five m.p.h. reduces fuel economy by an average of 6 percent.
- Avoid aggressive driving habits like rapid acceleration and braking, both of which reduce fuel economy. On average, an aggressive driver will consume an extra 125 gallons of gas annually.
- Travel light and pack smart. Place luggage and other items inside the car rather than on top or on the trunk to reduce wind resistance and increase mileage.
- Make sure your car is well maintained and running properly. If you do that, you can save up to 165 gallons of gas a year. Don't forget the tires, which should be properly inflated and aligned.
- Buy a hybrid vehicle or a much more fuel-efficient car. Choices include makes that can get up to sixty miles per gallon. We could cut global warming pollution by 10 percent if every American car buyer chose a vehicle that gets five more miles per gallon than the one they own now. Hydrogen or electrical-powered cars are the best of all current options.
- Drive less by combining trips when you have to run errands, by telecommuting once a week, by carpooling, and by using public transportation whenever possible. Each gallon of gasoline burned in a car emits twenty-five pounds of global warming pollution. We could save 5.85 billion gallons of oil and reduce carbon dioxide emissions by 143 billion pounds each year if every commuter in the U.S. telecommuted one day a week, according to Environmental Defense.

Buy Carbon Offsets

Concerned citizens can also help counteract their pollution by buying "offsets," meaning that you purchase credits in your name from certain organizations. If you bought ten tons of carbon offsets, for instance, the seller would guarantee that ten fewer tons of global warming pollution would be released into the atmosphere. The cost ranges from $4 to $8 per metric ton.

Without doing a good deal of research on where to buy offsets, you could become confused about how much pollution is actually being removed or reduced. The people at Environmental Defense have evaluated a number of offset organizations, including Carbonfund.org and Driving Green, to make shopping for these assets much easier.

You can review all of their recommended organizations at the web page found at:
www.fightglobalwarming.com/page.cfm?tagID=270

Even if you think you won't have much impact on the world's global warming problem, you really can make a difference. You can help and get started today by becoming a more energy-conscious consumer.

APPENDIX 6

HERE GOES THE NEIGHBORHOOD

TITAN, IN ALL THE SOLAR SYSTEM, IS CLOSEST IN GEOLOGICAL MAKE-UP TO THE EARTH

Titan, a moon of Saturn, is strikingly similar to Earth; it has rivers, rain, lakes, and glaciers. There may have even been life on Titan, but it now rains liquid methane and is 300 degrees below zero, but its atmosphere is believed to have been much more like ours before the composition of CO_2 and methane squeezed the oxygen level out of the atmosphere, leaving a thick hydrocarbon atmosphere that puts LA to shame. The atmospheric conditions that allow life to exist on a planet are a delicate combination of oxygen, CO_2, methane, and other gases. To tamper with that composition is to tamper with life itself.

APPENDIX 7

THE HOSTILE WORLD OF VENUS

Venus, our closest planetary neighbor, averages about 850 degrees Fahrenheit. The primary ingredient of the hostile, lifeless atmosphere is carbon dioxide, or CO_2; consequently, there is no limit to how hot the atmosphere can become.

As we approach a total of one billion automobiles in operation on Earth, all spewing into the air, perhaps we can visualize what the effects of having 3.5 billion automobiles on the road by the year 2030 will mean for our planet. Add to that 40 new coal-burning electric plants coming on line in the U.S. alone (eleven of them in Texas) and literally hundreds more all over the world (the present goal of the Chinese government is to fie up one new coal-burning electric plant there every week) and you get some idea of how critical and imperative it is for us to stop and change our direction and our use of fossil fuels before we pass the point of no return. The band width that supports life on Earth is between 20 and 120 degrees Fahrenheit. We should be very careful about disturbing the delicate balance in our atmosphere.

There are many today who feel we have missed our opportunity to correct the warming problem by 25 years of procrastination and debate, so they feel we must adapt.

But, in fact, can we adapt?

CAUTION

Caution in proceeding into uncharted waters is indeed warranted.

C. S. Goldsmith

METHANE
CO2
ECOCIDE

EPILOGUE

IN THE DEEPEST DEPTHS OF THE OCEAN LIES OUR FATE.

The world's scientific community, the Intergovernmental Panel on Climate Change (the "IPCC") has concluded that the world will be a minimum of 7 degrees, on average, warmer by the end of this century. We humans, the last of all hominids, are facing a far more precarious threat to our survival that in all of our human history, an ecological meltdown of natural environment, that will inevitably follow an 7 degree or more, on average, temperature increase.

But as bad as the radically changing climate may be for us and other species, there is a far more ominous threat that is camouflaged behind the thin veil of global warming, that has waited for millennia for temperatures to increase enough to set it free. Once unleashed, our chances of adapting to this warming trend, as many today would have us believe that we must do, are far less likely.

POSSIBLE SCENARIO

2024 First indication of measurable increases in methane released into the atmosphere (a 300% increase in methane released into the atmosphere than just five years ago).

2027 4% temperature spike now 7 degrees on average hotter, much faster than was anticipated or modeled on the computers. Methane is causing the increased temperature spikes.

2040 Massive heat wave causes crop failure; seasons are changing so quickly that farmers cannot react fast enough to calculate their responses to the climate change.

2052 Plants are withering and dying under the heat. Plant-eating animals - cattle, sheep, deer, are dying by the millions.

2060 3 % additional increase in temperature caused by massive methane release after Greenland and Antarctic ice sheets have melted, exposing the permafrost and releasing massive amounts of methane into the atmosphere.

2070 Oceanic collapse of major fish species due to loss of plankton and coral reef deterioration, overfishing, and dramatic changes in water temperatures that is affecting the breeding season of countless marine species.

2082 Regional nuclear war over dwindling natural resources.

2100 Methane release momentum is increasing, as warm oceanic water reaches deeper levels; the world is now, on average, 12 degrees warmer.

2150 66% of the world's population and 70% of all species may perish in the next 50 years.

The future for the remaining people and species on the planet remains in doubt, as the temperature increases are feeding upon themselves and stabilization of temperatures appears to be nowhere in sight,

We should be very cautious indeed in proceeding faster and further out onto the thin ice of global warming. We should instead marshal our scientific resources now to solve perhaps the single greatest threat to the survival of most of the species on the planet in the last 250 million years.

BIBLIOGRAPHY

Alley, R.S. & DeMenocal, P.B. (1998)
Abrupt climate changes revisited: How serious and how likely? Speeches delivered at the U.S. Global Change Research Program Seminar Series. February 23, 1998. Dirksen Senate Office Bldg., Washington, D.C.

Diamond, Jared, Penguin Press
Collapse

Gelbspan, Ross. (1997)
The heat is on: The high stakes battle over Earth's threatened climate. Reading, MA: Addison-Wesley.

Glick, D. & Hayden, T. (1997, December 8)
How to Beat the Heat. Newsweek, 34-38.

Max, Michael D. , Kluwis Academic Press
Natural Gas Hydrate in Oceanic and Permafrost Environments

Showstack, R. (1998).
Looking for El Nino's silver lining. EOS 79(8):2.

Stommel, H. & Stommel, E. (1979)
The Year without a Summer. Scientific American, 176-186.

Vogel, G. & Lawler, A. (1998, June 12)
Science, hot year, but cool response in Congress, P. 1684

Watson, Robert T.; Zinyowera, M. C., Moss, Richard H., and Dokken, David J. (1996). Climate change 1995: Impacts, adaptations and mitigation of climate change: Scientific-Technical Analyses. Cambridge, UK: Cambridge University Press.
Woodwell, G. M. (1989)

The warming of the industrialized middle latitudes 1985-2050: Causes and consequences. Climatic Change 15:31-50.

Information Unit for Conventions (IUC) United Nations Environment Programme (UNEP). Understanding Climate Change: A Beginner's Guide to the UN Framework Convention.